T0230428

BestMasters

Springer awards „BestMasters" to the best master's theses which have been completed at renowned universities in Germany, Austria, and Switzerland.

The studies received highest marks and were recommended for publication by supervisors. They address current issues from various fields of research in natural sciences, psychology, technology, and economics.

The series addresses practitioners as well as scientists and, in particular, offers guidance for early stage researchers.

Daniel Halwidl

Development of an Effusive Molecular Beam Apparatus

 **Springer** Spektrum

Daniel Halwidl
Wien, Österreich

BestMasters
ISBN 978-3-658-13535-5   ISBN 978-3-658-13536-2 (eBook)
DOI 10.1007/978-3-658-13536-2

Library of Congress Control Number: 2016935968

Springer Spektrum

Printed on acid-free paper

Springer Spektrum imprint is published by Springer Nature
The registered company is Springer Fachmedien Wiesbaden GmbH

Acknowledgements

I would like to express my gratitude and appreciation to my supervisors Prof. Ulrike Diebold and Jiri Pavelec, who supported me during my diploma thesis. I want to especially thank Jiri Pavelec for the technical drawings and encouraging me in my work. I also feel grateful to Prof. Michael Schmid, who was always available for fruitful discussions.

I want to thank Herbert Schmidt and Rainer Gärtner, who showed great skills and patience when it came to the production of the many parts the Molecular Beam consists of.

Many thanks I want to adress to Jan Hulva, Manfred Bickel, Florian Brunbauer and Jakub Piastek for their support in the laboratory.

Finally, I want to thank my family for supporting me during my study.

Daniel Halwidl

Abstract

This thesis describes the development of an effusive molecular beam apparatus, which allows the dosing of gases, liquids and solids. The apparatus was designed to adsorb precise and reproducible doses to a defined area on metal oxide samples, which is required in Thermal Programmed Desorption and other surface chemistry experiments in the "Machine for Reactivity Studies". The theoretical profile of the molecular beam has a core diameter of 3.5 mm and a standard core pressure of 4×10^{-8} mbar, while the background pressure of the beam is 4 orders of magnitude lower. The design and the construction of the apparatus is described. The calculated beam profile was experimentally confirmed and core pressures between 1.7×10^{-8} mbar and 2.9×10^{-6} mbar were measured.

Kurzfassung

Diese Diplomarbeit beschreibt die Entwicklung einer effusiven Molekularstrahl-apparatur für die Adsorption von Gasen, Flüssigkeiten und Feststoffen auf Metall-oxidproben in der "Machine for Reactivity Studies". Die Adsorption einer präzi-sen, reproduzierbaren Dosis auf einem wohldefinierten Bereich der Probenoberflä-che ist für Thermische Desorptionsspektroskopie und andere Methoden der Ober-flächenchemie notwendig. Das berechnete Profil des Molekularstrahls weist einen Kern mit Durchmesser 3.5 mm und einen Standarddruck von 4×10^{-8} mbar auf, während der Hintergrunddruck des Molekularstrahls um 4 Größenordnungen klei-ner ist. Der Entwurf und die Konstruktion des Apparats sind beschrieben. Das be-rechnete Profil wurde experimentell bestätigt und ein Druck von 1.7×10^{-8} mbar bis 2.9×10^{-6} mbar im Kern des Molekularstrahls gemessen.

Contents

List of Figures

List of Tables

1 Introduction

The "Machine for Reactivity Studies" (MRS) is a recently built vacuum chamber that will provide analysis techniques (Low Energy Electron Diffraction, Low Energy Ion Scattering, Temperature Programmed Desorption (TPD), Ultraviolet Photoelectron Spectroscopy, X-ray Photoelectron Spectroscopy) to investigate adsorption and surface chemistry at oxide surfaces. For cooling the sample a continuous-flow liquid He cryostat is used, which allows base sample temperatures below 50 K. Heating the sample to up to 1200 K is done by resistive heating of the sample holder.

Temperature Programmed Desorption (see Section 1.2) and other surface chemistry experiments require the adsorption of gases, evaporated liquids or sublimated solids on the sample. When desorbing the adsorbates again, usually only particles that desorb from the sample surface are of interest. One possibility is to dose to all surfaces in the camber and then come very close to the sample surface with a differentially pumped detector, where the entrance to the detector is a small aperture, only sampling species from the sample surface [9, 18]. That works fine for metal samples, which are directly spot-welded to thin heating wires, where the area of the heating wires is small compared to the sample area. Metal oxide samples that are used in the MRS can not be spot-welded but have to be clamped by a retaining frame. Since the frame is so close to the sample, the desorbing species would affect the TPD results. Therefore the adsorption has to be limited to the sample surface in the first place, which is achieved by the use of a molecular beam.

1.1 Molecular beams

A molecular beam (MB) is defined as a collision-free, spatially well-defined and directed flow of molecules [15]. To create a MB, gas with a certain temperature and pressure is expanded from a reservoir through a source nozzle into an evacuated chamber. The properties of the MB depend on the detailed form and size of the nozzle and the pressure and temperature of the gas in the reservoir. Once the initial MB has been created, it can be influenced further, for example by skimmers, apertures, shutters, and velocity selectors before hitting the sample.

Effusive and supersonic MBs are distinguished depending on the Knudsen number describing the expansion through the source nozzle. In supersonic beams the Knudsen number (see Section 2.1) is small: a gas with relatively high pressure (several atm.) reaches the speed of sound at the nozzle, and supersonic velocities in the following free jet expansion. Due to the frequent collisions during expansion the velocity distribution of the gas molecules is narrowed until, at certain distance from the source, the pressure in the beam drops to molecular flow conditions. In the *sudden freeze model* the surface at that distance is named the *quitting surface*, which then can be described as an effusive source. Therefore the effective radiating area can be larger than the geometrical area of the nozzle [3, 21]. The main advantages of a supersonic beam source are a narrow velocity distribution, variable kinetic energy, and a large degree of control over the internal energy. These features are essential in studies of the gas-surface dynamics, of surface scattering and diffraction or of fast transient kinetics [15].

In an effusive molecular beam, as in the present work, the Knudsen number in the source nozzle is large: the mean free path in the source is large compared to the size of the nozzle and particles don not collide with each other as the gas expands. Thus, the flow rate as well as the angular and velocity distribution of the effusing particles can be calculated exactly by gas kinetics [21]. If the thermal equilibrium in the source is not disturbed by the opening in the wall of the source, the beam contains a well defined equilibrium distribution of internal states and the same dimer fraction as that of the gas within the source. Sources for effusive molecular beams include thin-walled orifices, single tubes and capillary arrays. An effusive molecular beam apparatus using a capillary array as source is described by Libuda et al. [16].

1.2 Temperature programmed desorption

The desorption of adsorbed atoms or molecules can provide information concerning the strength of interactions between the surface and the adsorbed species [1]. In TPD a temperature ramp, linear in time, is applied to the sample and the rate of adsorbates desorbing into the gas phase is monitored by a detector. For preparing the adsorbate, the sample has to be sufficiently cold to adsorb molecules or atoms that impinge on the sample surface.

Desorption is an activated process with a rate constant k_d that obeys an Arrhenius dependence

$$k_d = A \exp\left(-\frac{E_d}{RT}\right),\qquad(1.1)$$

where E_d is the activation energy for desorption and A is a pre-exponential factor [1]. A has the unit s^{-1} and can be seen as an attempt frequency; for first order desorption (see below) A is of the same order of magnitude as the molecular vibrational frequency and usually assumed to be 10^{13} s^{-1}. With increasing temperature, thermal energy becomes sufficient to break surface bonds and desorption is observed. The rate of desorption may be formulated as

$$-\frac{dN}{dT} = N^m \frac{k_d}{\beta} = N^m \frac{A}{\beta} \exp\left(-\frac{E_d}{RT}\right),\qquad(1.2)$$

where N is the number of adsorbed molecules, β is the heating rate, and m is the order of the desorption. Although k_d increases exponentially with temperature, a maximum of the desorption rate is observed, because the surface coverage decreases during the desorption process.

Multilayer systems exhibit zero-order desorption where the rate of desorption does not depend on N. Here the rate of desorption increases exponentially with temperature and goes to zero when all adsorbates have been desorbed. With increasing initial coverage the multilayer desorption peak will grow in intensity and its maximum will shift to higher temperatures. The temperature of the leading edge of the desorption peak is independent of the initial coverage and depends only on the adsorbate; the influence of the surface is almost completely screened. The bonds in the second and subsequent monolayers are usually weaker than the bond of the first monolayer to the substrate, therefore multilayer desorption peaks appear at lower temperatures relative to the desorption peaks from the first monolayer.

Desorption from monolayer or submonolayer coverages is in the simplest case of first order. The rate of desorption increases exponentially with temperature, reaches a maximum and decreases as the number of possible desorption sites (the adsorbed molecules) decreases. With increasing initial coverages the symmetric desorption peak grows in intensity and its maximum should stay at the same temperature. However, in practice it is often observed that the maximum shifts to lower or higher temperature with increasing initial coverage as lateral interactions between the molecules influence the activation energy for desorption.

2 Flow of gases

When planning a molecular beam apparatus the knowledge of the different flow regimes is important . These play a key role in the formation of molecular beams and in the design of conductance requirements in the whole vacuum system.

2.1 Flow regimes

The mean free path, λ, is defined as the average distance a particle in a gas travels before colliding with another particle. A way to obtain the mean free path is to calculate it via the viscosity, η, as [24, p.664]

$$\lambda = \frac{\eta \cdot \sqrt{\pi}}{2 \cdot p} \cdot \sqrt{\frac{2k_B T}{m}} = \frac{\eta \cdot \pi}{4 \cdot p} \cdot \bar{v} , \qquad (2.1)$$

where $\bar{v}$ is the average velocity of the gas particles, given by

$$\bar{v} = \sqrt{\frac{8k_B T}{\pi m}} . \qquad (2.2)$$

Three idealized regimes are distinguished depending on the pressure and the geometry of the vacuum equipment the gas is flowing through (e.g. the diameter of a tube or orifice, the length of a channel) : molecular flow, continuum flow and transition flow [20, pp.25–27].

If the pressure is sufficiently low, the mean free path of the particles will be much greater than the diameter of the vacuum equipment. Therefore the majority of particles will move along straight trajectories until hitting a wall. Collisions between the particles occur very rarely, they move independently of each other. These conditions are called molecular flow. The flow is only caused by the kinetic energy of the particles.

If the pressure is sufficiently high, the mean free path of the particles is much smaller than the diameter or length of the vacuum equipment. Therefore particles will collide very often with each other, resulting in frequent exchange of momentum and energy. The particles can be treated as a continuum and flow is caused by local pressure gradients. Thus, this regime is named continuum flow. Continuum flow can be either turbulent or laminar viscous.

Transition flow occurs when the pressure is between the limits above; neither molecular or continuum flow prevails.

To quantitatively distinguish the flow regimes the dimensionless Knudsen number, Kn, is defined as

$$Kn = \frac{\lambda}{D} \, , \tag{2.3}$$

where λ is the mean free path and D is the characteristic dimension of the vacuum equipment (e.g. diameter of a tube). The transition between flow regimes is continuous, but a classification by the Knudsen number is:

$$
\begin{aligned}
Kn &> 1 \quad \text{molecular flow,} \\
1 > Kn &> 0.01 \quad \text{transition flow,} \\
Kn &< 0.01 \quad \text{continuum flow.}
\end{aligned}
\tag{2.4}
$$

2.2 Conductance

In the following sections the conductance in the different flow regimes is discussed. The conductance, C, of a tubing component is generally defined as [13, p.92]

$$C = \frac{q_{pV}}{p_1 - p_2} \, , \quad [C] = \mathrm{m}^3/\mathrm{s} \, , \tag{2.5}$$

where q_{pV} is the throughput and $p_1 - p_2$ is the pressure difference between the entrance and the exit of the tubing component.

2.2.1 Molecular flow

Particles that enter a tube will move on a straight trajectory until they hit either the wall of the tube or make it directly to the exit of the tube. Particles that have collided with the wall will be scattered in random directions and thus go back and forth in the tube until they leave the tube either through the entrance or the exit area. Therefore the conductance in molecular flow, C_{mol}, can be described statistically by the conductance of the entrance area, C_O, and the transmission probability, P, for a particle to make it through the tubing component [13, pp.135–139]:

$$C_{mol} = C_O \cdot P . \tag{2.6}$$

Orifice

The conductance, C_O, of an ideal orifice with infinitely small wall thickness ($L = 0$) and area A is

$$C_O = \frac{\bar{v}}{4} A . \tag{2.7}$$

The transmission probability of an ideal orifice ($L = 0$) is obviously 1. Table 2.1 shows the orifice conductance per area, C_O/A, for different gases according to Eq. (2.7).

Tube

The transmission probability, P_T, for a tube with length L and diameter D is given by [13, pp.142–144]

$$P_T(L, D) = \frac{14 + 4\frac{L}{D}}{14 + 18\frac{L}{D} + 3\left(\frac{L}{D}\right)^2} . \tag{2.8}$$

Gas	$\bar{v}$ [m/s]	C_O/A [l/(s · cm^2)]
H$_2$	1762	44.0
He	1246	31.1
H$_2$O	587	14.7
N$_2$	471	11.8
Ar	394	9.8
Xe	218	5.4

Table 2.1: Molecular orifice conductance per area for gases at $20\,^\circ C$.

For a short tube with $L << D$ the transmission probability P_T becomes

$$P_{T,\text{short}} = 1 - \frac{L}{D}\,, \tag{2.9}$$

and for a long tube with $L >> D$

$$P_{T,\text{long}} = \frac{4D}{3L}\,. \tag{2.10}$$

Figure 2.1 shows the transmission probability, P_T, according to Eq. (2.8). The conductance of a tube, C_T, is then given by Eq. (2.6).

$$C_T = C_{0,T} P_T\,, \tag{2.11}$$

where $C_{0,T}$ is the orifice conductance of the entrance area of the tube.

Tube with annular cross section

Several formulas for the conductance of a tube with an annular (ring-like) cross section can be found in the literature, see Appendix B.1, Page 93. The formula, which will be used in this work to calculate the conductance, C_{ann}, of an annular tube (important for the pumping speeds in the pumping stages of the MB, see Section 4.4.6, Page 68) with outer diameter D_o, inner diameter D_i and length L is

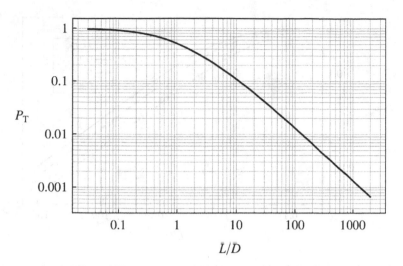

Figure 2.1: Transmission probability, P_T, for a tube as a function of the relative length, L/D, according to Eq. (2.8).

given by

$$C_{\text{ann}} = 3.81 \sqrt{\frac{T}{M}} \frac{(D_o - D_i)^2 (D_o + D_i)}{L + 1.33(D_o - D_i)} \,, \tag{2.12}$$

where M is the relative molecular mass; dimensions are in cm and C_{ann} in l/s. Figure 2.2 shows the transmission probability, P_{ann}, obtained from Eq. (2.12).

Tubing components in series

The simplest way to calculate the conductance of composite systems is the electric circuit analogy [13, pp.91–94]. If n elements are connected parallel, the conductances, C_i, add up to give

$$C = C_1 + C_2 + \dots + C_n \,. \tag{2.13}$$

and if connected in series, the inverse conductances add up to give

$$\frac{1}{C} = \frac{1}{C_1} + \frac{1}{C_2} + \dots + \frac{1}{C_n} \,. \tag{2.14}$$

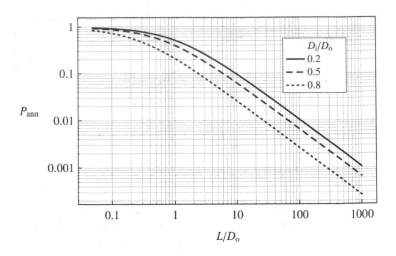

Figure 2.2: Transmission probability for annular tubes, P_{ann}, with different inner to outer diameter ratio, D_i/D_o, as a function of the relative length, L/D_o, obtained from Eq. (2.12). The transmission probability was obtained by dividing the conductance, C_{ann}, by the orifice conductance of the annular cross section.

The latter case also applies to the calculation of the effective pumping speed at a certain point in a vacuum system. It is given by the pumping speed, S_p, provided by the vacuum pump and the conductance, C, of the tubing connecting said point to the pump. The effective pumping speed, S_{eff}, is then given by

$$\frac{1}{S_{eff}} = \frac{1}{S_p} + \frac{1}{C}. \tag{2.15}$$

However, two conditions must be met for the circuit analogy to hold: the rate of flow of gas through a single tube must be proportional to the pressure difference between its ends, and the conductivity of a tube must be independent of the components to which its ends are connected. The latter is in general not true, as Eq. (2.14) gives a rather poor approximation with errors up to 40 % for two tubes with identical radii connected in series [19]. A better expression for the transmission probability, P, for the connected tubes is given by

$$\frac{1}{P} = \frac{1}{P_1} + \frac{1}{P_2} - 1 , \tag{2.16}$$

where P_1 and P_2 are the transmission probabilities of the respective tubes.

An advanced version of Eq. (2.16) is also considering changes of cross sections and gives the transmission probability, P_{1n}, of n components as [20, p.47]

$$\frac{1}{A_1} \left(\frac{1}{P_{1n}} - 1 \right) = \sum_{i=1}^{n} \frac{1}{A_i} \cdot \left(\frac{1}{P_i} - 1 \right) + \sum_{i=1}^{n-1} \left(\frac{1}{A_{i+1}} - \frac{1}{A_i} \right) \cdot \delta_{i,i+1} , \tag{2.17}$$

where

P_i transmission probability of the i-th component,

A_i cross section of the i-th component,

$\delta_{i,i+1}$ $= 1$, if $A_{i+1} < A_i$ (reduction of cross section),

           $= 0$, if $A_{i+1} \geq A_i$ (no reduction of cross section).

The conductance of n components, C_{1n}, is then calculated by multiplying the transmission probability, P_{1n}, with the orifice conductance of the cross section of the first component, C_{0,A_1}, giving

$$C_{1n} = C_{0,A_1} P_{1n} . \tag{2.18}$$

2.2.2 Continuum flow

In continuum flow the conductance is pressure dependent [13, pp.96–103]. For a converging nozzle the throughput, q_{pV}, follows from the conservation of the mass flowing from the upstream to the downstream side:

$$q_{pV} = A_{\min} \cdot \sqrt{\frac{\pi}{4}} \cdot p_1 \cdot \bar{v} \cdot \psi \left(\frac{p_2}{p_1} \right) , \tag{2.19}$$

Gas species	Isentropic exponent κ [13, p.97]	Critical pressure ratio p^*/p_1	Maximum value of flow function $\psi_{critical}$
Inert gases	≈ 1.667	0.487	0.513
Diatomar gases	$= 1.4$	0.528	0.484
Complicated molecules (e.g. oil vapours)	≈ 1.1	0.585	0.444

Table 2.2: Isentropic exponent, critical pressure ratio, and maximum value of the flow function for various gas species.

where p_1 and p_2 are the pressures at the upstream and downstream side, respectively and A_{min} is the narrowest cross section of the nozzle. The *flow function* ψ describes the dependence on the pressure ratio and is given by

$$\psi\left(\frac{p_2}{p_1}\right) = \sqrt{\frac{\kappa}{\kappa - 1} \cdot \left(\left(\frac{p_2}{p_1}\right)^{\frac{2}{\kappa}} - \left(\frac{p_2}{p_1}\right)^{\frac{1+\kappa}{\kappa}}\right)}, \tag{2.20}$$

where κ is the isentropic exponent[1] of the gas. Figure 2.3a shows that ψ has a maximum, $\psi_{critical}$, at the critical pressure, p^*, when the gas flow reaches the speed of sound between $p_2 = 0.45p_1$ and $0.6p_1$ (depending on κ). A further lowering of the downstream pressure would decrease the mass flow again, according to Eq. (2.20). However, in reality the flow function takes on its maximum value, $\psi_{critical}$, once the gas flow reaches the speed of sound as no further interaction with the molecules upstream is possible, see Fig. 2.3b; the flow condition is called choked flow. Table 2.2 shows the isentropic coefficient, the critical pressure ratio and the maximum value of the flow function for various gas species. The expressions for $\psi_{critical}$ and p^*/p_1 are given in Eqs. (2.21) and (2.22), respectively.

$$\psi_{critical} = \psi\left(\frac{p^*}{p_1}\right) = \left(\frac{2}{\kappa + 1}\right)^{\frac{1}{\kappa - 1}} \cdot \sqrt{\frac{\kappa}{\kappa + 1}} \cdot \tag{2.21}$$

$$\frac{p^*}{p_1} = \left(\frac{2}{\kappa + 1}\right)^{\frac{\kappa}{\kappa - 1}} \tag{2.22}$$

---

[1] the ratio of the heat capacity at constant pressure to the heat capacity at constant volume

Orifice

For a thin walled orifice with cross section A_O the expression for the throughput of the converging nozzle, Eq. (2.19), has to be corrected for the sudden contraction of the flow at the orifice opening. Depending on the pressure ratio, the following correction for A_{min} is applied [13, p.103]:

$$A_{min} = \begin{cases} 0.60 \cdot A_O & \text{if } p_2 \approx p_1 \\ 0.86 \cdot A_O & \text{if } p_2 < p^* \quad \text{(choked flow)}. \end{cases} \quad (2.23)$$

Therefore the conductance of an orifice in continuum flow under chocked flow conditions is given by

$$C_{O,cont} = 0.86 \cdot A_O \cdot \sqrt{\frac{\pi}{4}} \cdot \bar{v} \cdot \psi_{critical}. \quad (2.24)$$

Note that the conductance of the orifice in chocked flow is independent of the pressure, as in molecular flow.

Tube

The type of continuum flow in a tube is characterised by the Reynolds number [13, pp.116–119]:

$$Re = \frac{32}{\pi^2} \cdot \frac{q_{pV}}{\eta \cdot \bar{v} \cdot D} \begin{cases} < 2300 & \text{laminar viscous flow} \\ > 4000 & \text{turbulent flow}. \end{cases} \quad (2.25)$$

For a tube connecting two vessels with pressures p_1 and p_2, the conductance in laminar viscous flow, $C_{T,lam,}$, is given by

$$C_{T,lam} = \frac{\pi}{256} \cdot \frac{1}{\eta} \cdot \frac{D^4}{L} \cdot (p_1 + p_2), \quad (2.26)$$

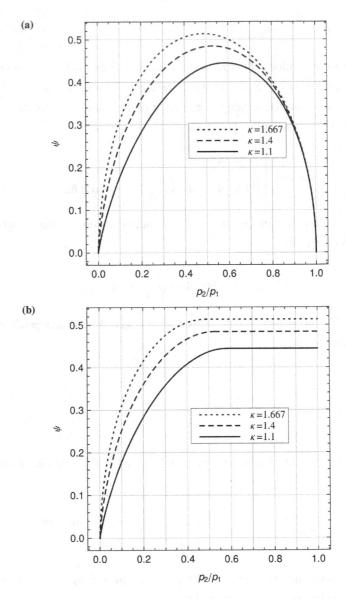

Figure 2.3: (a) Theoretical behaviour of the flow function, ψ, according to Eq. (2.20). (b) Real behaviour of ψ according to Eqs. (2.20) and (2.21). As the critical pressure is reached, the flow function stays on its critical value, ψ_{critical}.

and in turbulent flow, $C_{\text{T,turb}}$, by

$$C_{\text{T,turb}} = 1.015 \cdot D^{\frac{19}{7}} \cdot \left(\frac{\bar{v}^6}{\eta}\right)^{\frac{1}{7}} \cdot \left(\frac{p_1 + p_2}{L}\right)^{\frac{4}{7}} \cdot (p_1 - p_2)^{-\frac{3}{7}}, \qquad (2.27)$$

where η is the viscosity of the gas.

2.2.3 Transition flow and conductance over the whole pressure range

When conditions change from molecular to continuum flow, both types of flow occur at the same time. The resulting flow is called transition flow, which is difficult to describe exactly, but good experimental data and approximations are available.

Orifice

The conductance of an orifice is given by Eqs. (2.7) and (2.24) for molecular and continuum flow, respectively. If in continuum flow one assumes choked flow and $\psi_{\text{critical}} = 0.484$ (for air) then the ratio of the conductance in the continuum limit, $C_{\text{O,cont}}$, to the molecular conductance, C_{O}, is

$$\frac{C_{\text{O,cont}}}{C_{\text{O}}} = 0.86 \cdot \psi_{\text{critical}} \cdot 2\sqrt{\pi} = 1.48 . \qquad (2.28)$$

Therefore the conductance of the orifice increases by a factor of 1.48 when changing from molecular to continuum flow.

Figure 2.4 shows the measured conductance, C, of an orifice, normalized to its molecular conductance, over a wide range of the inverse Knudsen number. In the continuum limit (i.e. $Kn^{-1} \approx 10000$) the experimental data is in good agreement with Eq. (2.28). The observed maximum at $Kn^{-1} \approx 100$ is explained by the not yet fully developed clogging of the flow at the downstream side of the orifice [12].

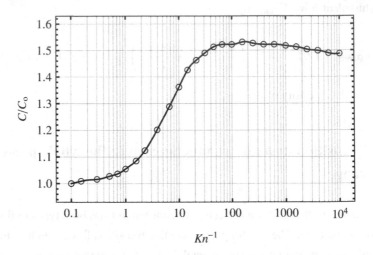

Figure 2.4: Interpolation (line) of the experimental data (circles) for the ratio C/C_O as a function of the inverse Knudsen number from [12, Fig.2]. Values for the interpolation are listed in Table B.3, Page 99.

Therefore the ratio of the orifice conductance, C_{orifice}, to the molecular orifice conductance, C_O, over the whole pressure range can be expressed by

$$\gamma(Kn) = \frac{C_{\text{orifice}}}{C_O} = \begin{cases} 1 & \text{if} \quad Kn^{-1} < 0.1, \\ \epsilon & \text{if} \quad 0.1 < Kn^{-1} < 10000, \\ 1.48 & \text{if} \quad Kn^{-1} > 10000, \end{cases} \qquad (2.29)$$

where ϵ is the interpolated ratio from Fig. 2.4.

The conductance of an orifice, C_{orifice}, for air over the whole pressure range is therefore given by

$$C_{\text{orifice}} = \gamma(Kn)C_O. \qquad (2.30)$$

Tube

A long tube can be approximated by two parallel tubes of which one is considered to be in molecular and the other in laminar viscous flow [13, pp.159–162]:

$$C_T \approx C_{T,mol} + C_{T,lam} . \tag{2.31}$$

This equation correctly describes the conductances in the molecular and continuum limit and is an approximation in transition flow. By adding the respective conductances from Eqs. (2.10) and (2.26) one obtains

$$C_T = \frac{\pi}{12} \left(\frac{3}{32} \cdot \frac{\bar{p} \cdot D}{\eta \cdot \bar{v}} + Z \right) \cdot \frac{D^3}{L} \cdot \bar{v} , \tag{2.32}$$

where $\bar{p}$ is the average of the pressures at the entrance and the exit of the tube. According to the addition, Z equals 1. However, *Knudsen* found a semi empirical expression for the dimensionless correction factor given by

$$Z = \frac{1 + \frac{1.28}{Kn}}{1 + \frac{1.58}{Kn}} , \tag{2.33}$$

where the Knudsen number is to be calculated at the average pressure $\bar{p}$. The dimensionless bracket term in Eq. (2.32) is called the *conductance function*, f, and is only dependent on the Knudsen number:

$$f = \frac{3}{32} \cdot \frac{\bar{p} \cdot D}{\eta \cdot \bar{v}} + Z = \frac{3 \cdot \pi}{128} \cdot \frac{1}{Kn} + Z . \tag{2.34}$$

Figure 2.5 shows f as a function of the inverse Knudsen number. It is 1 in the molecular region, decreases with increasing pressure to a minimum at $Kn^{-1} \approx 0.6$ and then increases rapidly. The conductance, C_T, of a tube with diameter D and length L over the whole pressure range is therefore expressed by

$$C_T = \frac{\pi}{12} \cdot f \cdot \frac{D^3}{L} \cdot \bar{v} . \tag{2.35}$$

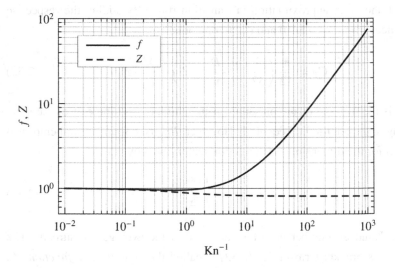

Figure 2.5: *Conductance function*, f, and correction factor, Z, from Eqs. (2.33) and (2.34) as a function of the inverse Knudsen number.

3 Effusive molecular beam sources

3.1 Thin-walled orifice

The effusive flow of gas from a thin-walled reservoir is considered. The gas is assumed to be Maxwellian (see Appendix A.1, Page 91), thus having the average particle velocity

$$\bar{v} = \sqrt{\frac{8k_B T}{\pi m}}, \tag{3.1}$$

where T is the absolute reservoir gas temperature. The number of particles that leave the source per unit time into the solid angle element $d\Omega$ can be written as

$$I(\theta) \, d\Omega = \frac{n_r \bar{v}}{4} d\sigma \, \cos(\theta) \frac{d\Omega}{\pi}, \tag{3.2}$$

where n_r is the number density in the reservoir, $d\sigma$ is an element of the orifice area and θ is the angle between $d\Omega$ and the normal of the orifice plane [23]. The intensity at an arbitrary point downstream of the orifice follows from integrating Eq. (3.2) over the whole area of the orifice. The integral can be solved analytically for any point in case of a rectangular orifice and for points on the centerline in case of a circular orifice [10].

For distances which are large compared to the orifice dimensions, the intensity follows from Eq. (3.2) by replacing $d\sigma$ by the orifice area σ. The intensity in the forward direction ($\cos(\theta) = 1$) is then

$$I(0) = \frac{n_r \bar{v} \sigma}{4\pi} = \frac{p_r}{k_B T} \frac{\bar{v} \sigma}{4\pi}, \quad [I(0)] = \text{particles sr}^{-1}\text{s}^{-1}, \tag{3.3}$$

where p_r is the reservoir pressure. The angular dependent intensity, $I(\theta)$, is given by

$$I(\theta) = I(0)\cos(\theta),\qquad(3.4)$$

which is why a thin-walled orifice is also called a *cosine emitter*. However, as mentioned in Ref. [14], from a practical point of view the intensity per unit solid angle is of less interest than the intensity per unit area of the sample, denoted $G(\theta)$. For a planar sample mounted perpendicular to the centerline of any emitter with angular intensity $I(\theta)$ the intensity per unit area, $G(\theta)$, is given by [14]

$$G(\theta) = I(\theta)\cos^3(\theta),\qquad(3.5)$$

and therefore in the case of a thin-walled orifice by

$$G_{\text{orifice}}(\theta) = I(0)\cos^4(\theta).\qquad(3.6)$$

Fig. 3.1 shows the $\cos^4(\theta)$ dependency of G_{orifice}.

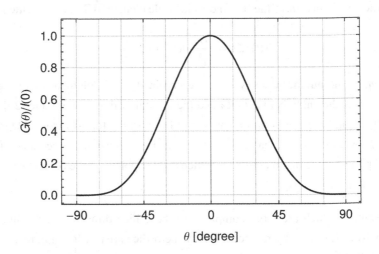

Figure 3.1: Intensity per unit area, $G_{\text{orifice}}(\theta)$, in a plane perpendicular to the centerline of a thin-walled orifice. The intensity was normalized to the intensity in forward direction.

3.2 Single tube

A tube with length L and inner radius R, which is connected to a gas reservoir with pressure p_r is considered. The Knudsen number with respect to the tube length is assumed to be $Kn_L \geq 1$. The pressure at the exit should be negligibly low.

Downstream from the tube two contributions to the flux of particles have to be considered: particles that come directly from the entrance of the tube without having collided with the wall and particles that have collided with the wall. For the latter a diffuse, i.e. cosine distributed, scattering from the walls is assumed.

Clausing [7] gave one of the first derivations for the flow from a single tube under the stated conditions. A similar derivation is given in Ref. [21], from which the main results will be stated. The particle flux, $I(\theta)$, leaving the tube exit under an angle θ is given by

$$I(\theta)\mathrm{d}\Omega = \frac{n_r \bar{v} A}{4\pi} j(\theta)\mathrm{d}\Omega \,, \tag{3.7}$$

where $A = \pi R^2$ is the cross section of the tube and $j(\theta)$ is the angular distribution of the particle flux. Two angular regions for the particle flux leaving the exit of the tube are distinguished:

$$\tan(\theta) \begin{cases} < 2R/L & \text{contributions directly from the reservoir and from} \\ & \text{the wall} \\ \geq 2R/L & \text{contributions from the wall only} \end{cases} \tag{3.8}$$

The angular distribution, $j(\theta)$, for $\tan(\theta) < 2R/L$ is given by

$$j(\theta) = \alpha \cos(\theta) + \frac{2}{\pi} \cos(\theta)$$
$$\cdot \left[(1-\alpha)\left(\arccos(q) - q\sqrt{1-q^2}\right) + \frac{2(1-2\alpha)}{3q}\left(1 - \sqrt{(1-q^2)^3}\right) \right] \tag{3.9}$$

and for $\tan\theta \geq 2R/L$ by

$$j(\theta) = cos(\theta)\left[\alpha + \frac{4(1-2\alpha)}{3\pi q}\right],\qquad(3.10)$$

where $q = \frac{L}{2R}\tan(\theta)$, and α is a term describing the wall collision rate.[1] Figure 3.2 shows the angular distribution, $j(\theta)$, for tubes with different length to diameter ratios.

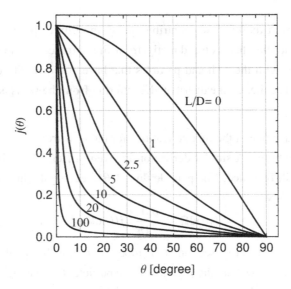

Figure 3.2: Angular intensity distribution, $j(\theta)$, of a tube according to Eqs. (3.9) and (3.10). The orifice limit, $L/D = 0$, and tubes with different length to diameter ratios are shown.

3.3 Capillary array

Single tubes have the disadvantage that the pressure in the reservoir has to be kept rather low to meet the condition for molecular flow with respect to the length of

---

[1]$\alpha = \frac{1}{2} - \frac{1}{3\beta^2}\left[\frac{1-2\beta^3+(2\beta^2-1)\sqrt{1+\beta^2}}{\sqrt{1+\beta^2}-\beta^2\left[\ln\left(\sqrt{1+\beta^2}+1\right)-\ln\beta\right]}\right],\quad \beta = \frac{2R}{L}$

the tube. By combining many parallel single tubes in a capillary array the intensity can be vastly increased without losing the directivity of the flux [21].

Figure 3.3 shows an example for the intensity distribution of a capillary array with diameter 1 mm, consisting of 1795 tubes ($L/D = 20$); the distance to the sample is 90 mm. The distribution was calculated by discretely summing up the contribution of each tube, $j(\theta)$, according to Eqs. (3.9) and (3.10). The calculated intensity distribution shows a qualitatively good agreement with calculations in Ref. [6].

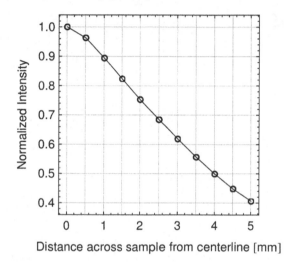

Figure 3.3: Intensity distribution of a capillary array on a planar sample, normalized to the centerline intensity. The capillary array, diameter 1 mm, consists of 1795 tubes ($L/D = 20$), the distance to the sample is 90 mm.

4 Molecular Beam

4.1 Requirements

Profile

Temperature Programmed Desorption requires the even adsorption of gas on the sample, while on surfaces next to the sample, e.g. the sample holder frame and the heating wires, as little gas as possible should adsorb. Hence the ideal intensity profile of the Molecular Beam on the sample is a top-hat profile, as shown in Fig. 4.1. The MB should have a constant core intensity over an area with diameter 3.5 mm, a size suitable for most of the samples to be used in the MRS. Outside of the core area the intensity should abruptly drop to zero.

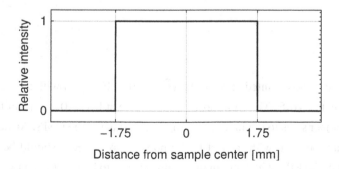

Figure 4.1: Ideal intensity distribution of the MB on the sample.

In a real molecular beam profile the intensity will not be perfectly constant and the edge will have a finite dimension over which the intensity will drop to a certain background intensity. The intensity in the core area should be constant within 1 %. The limit for the dose originating from the edge is set to 2.5 % of the dose in the core area. The background intensity should be kept as low as possible.

Dose rate and species to be dosed

Doses as small as 0.01 Langmuir (L) with a reproducibility better than 1 % should be achieved. The highest dose rate should be 5 L/s, which corresponds roughly to the highest pressures used when introducing gas directly into the chamber through a leak valve, i.e. using "background" deposition.

Dosing of gases, liquids (H_2O) and evaporable solids should be possible. Gases to be dosed include, among others, the noble gases (except for Rn), H_2, CO, CO_2, and O_2. The dosing of a mixture of two gases, or a gas and a liquid, should be possible. When dosing a mixture of two species, the amount of any potential reaction products in the MB should be minimized.

Design

The MB has to be mounted on a ConFlat® 100 (CF 100) port with a port length of 235 mm on the MRS. The frontmost part has to be at least 30 mm from the center of the chamber so as to not interfere with the cryostat and the QMS. An adjustment of the beam position with respect to the port axis of ±3 mm should be possible. The parts of the MB, which are in the chamber have to be non magnetic to prevent disturbance of LEED and UPS measurements.

For differential pumping of the MB two turbo molecular pumps (TMP) with a pumping speed of 260 l/s for N_2 are available.

4.2 Concept

Working principle of the Molecular Beam

Figure 4.2 shows a schematic of the working principle of the MB. Gas effuses from the source. A part of the gas is used to form the beam by placing a beam defining aperture between source and sample, hereafter called Aperture 2. The area on the sample from which the whole source is seen (through Aperture 2) is called the MB core, the area on the sample from which only a part of the source is seen is called the MB penumbra.

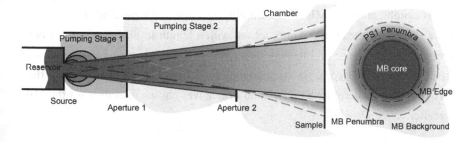

Figure 4.2: Working principle of the MB. Directed gas from the Reservoir is colored blue, background gas is colored gray. On the sample the MB core, the MB penumbra, the PS1 penumbra and the MB background from PS2 are shown.

To reduce the MB background pressure an additional aperture, called Aperture 1, is placed between the source and Aperture 2 to differentially pump the MB; Aperture 1 is barely larger than it would have to be from a geometric point of view, so as not to disturb the passing MB. The volume between source and Aperture 1 is called Pumping Stage 1 (PS1), the volume between Aperture 1 and Aperture 2 is called Pumping Stage 2 (PS2). The background gas in PS1 effuses through Aperture 1 into PS2, and a part effuses directly through Aperture 2 to the sample. The area on the sample from which only a part of Aperture 1 can be seen (through Aperture 2) is called the PS1 penumbra. The MB edge consists of the MB penumbra and the PS1 penumbra.

Source type

Since any source is finite in size, the area with full intensity of the MB on the sample will not have an infinitely sharp edge. To achieve a profile as close as possible to the ideal top-hat profile with a very small penumbra, the source dimension has to be very small. That rules out the use of a capillary array, as it relies on many tubes which are spread over the array.

Another possibility would be a single tube, which would concentrate the intensity in forward direction depending on its length to diameter ratio. While that behaviour is favourable when considering the gas load to the pumping stages, it brings a problem for the MB core: assuming a source to sample distance of 90 mm and a sample diameter of 3.5 mm, a tube with a length to diameter ratio of only 1 would already cause an intensity difference of 1.41 % between the center and the edge of the MB core. In comparison, an ideal orifice causes only a difference of 0.08 %. Therefore a thin-walled orifice was chosen as source for the MB, hereafter called *Orifice*.

Shutter

Precise doses are achieved by opening and closing a mechanical shutter for a defined time period. The Shutter is located in PS2 immediately after the exit of Aperture 1. By closing the shutter the direct beam is interrupted and the pressure in PS2 lowered by reducing the conductance between the pumping stages.

The mount for an additional shutter between Orifice and Aperture 1 is prepared, in case the shutter in PS2 is not sufficient to hinder gas from the direct beam from reaching the sample. The concept for this additional shutter is a piezoelectric plate bender, which moves a thin foil glued to its end into the line of sight from the Orifice to Aperture 1.

Reservoir

Species to be dosed are leaked into the Reservoir either by a manual leak valve or two controlled piezoelectric leak valves (Piezo leak valve). Figure 4.3 shows a schematic of the Reservoir supply and the pumping of the MB.

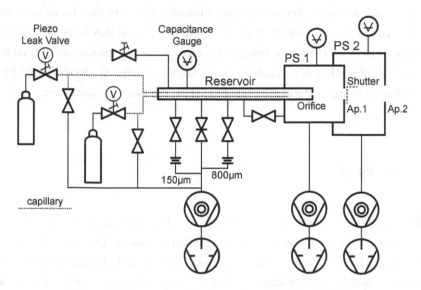

Figure 4.3: Reservoir supply and pumping schematic of the MB.

The two Piezo leak valves are connected via capillaries to the Reservoir, the manual leak valve is directly connected. The capacitance manometer measures the Reservoir pressure independently of the gas species and provides the feedback for the controlled Piezo leak valve. The Reservoir is connected to a TMP by a CF 40 gate valve and two pumping orifices with 150 μm and 800 μm diameter bypassing the gate valve. During operation of the MB the gate valve is closed and the Reservoir is pumped through one of the pumping orifices for easier control of the Reservoir pressure. After the MB operation the gate valve is opened to pump the Reservoir. The Reservoir and PS1 can be connected by a valve to protect the thin-walled Orifice from too high pressure differences while venting or pumping down the system.

To mix two gases in the Reservoir, each gas is admitted to the Reservoir through its own capillary. The capillaries end in the foremost section of the Reservoir, close to the Orifice. The capillaries are used for two reasons. First, the relatively high pressure at the entrance and the narrow diameter of the capillary ensure continuum flow which should prevent any flow back from the mixture to the gas bottle, keeping the gas clean. Second, the gases intermix closer to the Orifice and thus will make less contact with the walls of the Reservoir until they leave through the Orifice. If any reactions of the two gases occur at the walls, the amount of reaction products in the MB will be minimized. The capillaries and the Reservoir will be heated when using evaporated liquids and solids to prevent condensation.

4.3 Theory

The MB concept includes a circular source and two circular apertures to form a beam which is hitting the sample. Before the dimensions and properties of the MB are calculated, the general system source–aperture–sample (plane) is discussed, since its properties are important for the calculations of the doses delivered by the MB on the sample.

4.3.1 Core and Penumbra

A circular source with diameter d and a circular aperture with diameter D are assumed to be coaxially aligned, and separated by the distance x, see Fig. 4.4. In a plane of distance y from the aperture, perpendicular to the axis source–aperture, a circular area exists from which the unobscured source can be seen through the aperture. That area is called *core* and its radius the *core radius* r_c. Around the core a ring-shaped area exists in which the source is partially obscured by the aperture. That area is called *penumbra* and its ring width the *penumbra width w*.

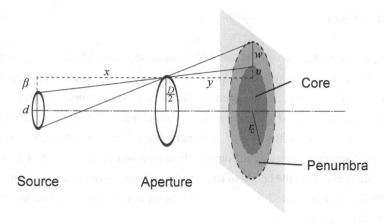

Figure 4.4: Geometry of source, aperture and plane with core and penumbra.

Penumbra width

The penumbra width is only dependent on the source diameter, the distance from source to aperture and the distance from aperture to the plane as shown in the following. One finds the following relations from Fig. 4.4:

$$\frac{w + v}{y} = \frac{d + \beta}{x} \tag{4.1}$$

$$\frac{v}{y} = \frac{\beta}{x} \quad \Rightarrow \quad v = y\frac{\beta}{x} \tag{4.2}$$

Inserting the relation for v, Eq. (4.2), in the first relation, Eq. (4.1), yields

$$\frac{w}{y} + \frac{\beta}{x} = \frac{d}{x} + \frac{\beta}{x}, \quad \Rightarrow \quad w = d\frac{y}{x}, \tag{4.3}$$

showing that w is directly proportional to d and y, and indirectly proportional to x. The penumbra width is *not* dependent on the core radius or the aperture diameter.

Penumbra dose

In the penumbra the intensity is steadily decreasing when radially going from its inner edge, $r = r_c$, to the outer edge, $r = r_c + w$, as the source is increasingly obscured by the aperture. If the aperture diameter is large compared to the source diameter, the unobscured source can be approximated by a circular segment with height h. At a plane radius of $r = r_c$ (inner edge of the penumbra) the height of the circular segment is d (source completely unobscured) and at $r = r_c + w$ (outer edge of the penumbra) the height of the circular segment is zero (source completely obscured). Therefore the height, h, of the circular segment as a function of the plane radius, r, is given by

$$h(r) = d \left(1 - \frac{r - r_c}{w}\right) . \tag{4.4}$$

The unobscured source area in the penumbra, $A_{s,p}(r)$, is therefore given by the area of a circular segment[1], A_{cs}, of radius $d/2$ and height $h(r)$ from Eq. (4.4):

$$A_{s,p}(r) = A_{cs}(h(r), d/2) . \tag{4.5}$$

Since the unobscured source area is equivalent to the intensity, I, in the penumbra one can write

$$I(r) = A_{s,p}(r) \tag{4.6}$$

Figure 4.5 shows the intensity in the penumbra according to Eq. (4.6), where the intensity in the core, I_c, and thus at $r = r_c$ is assumed to be 1.

Since $I(r)$ decreases almost linearly, it was assumed that the dose to the penumbra, D_p, could be approximated by a dose equivalent to the full intensity ($I = I_c = 1$) on half the penumbra area, as long as the ratio of penumbra width to core radius is small. To verify this assumption, the dose D_p, obtained by integrating $I(r)$

---

[1] The area of an arbitrary circular segment, A_{cs}, with radius r' and height h' is given by

$$A_{cs}(h', r') = r'^2 \arccos\left(1 - \frac{h'}{r'}\right) - \sqrt{2r'h' - h'^2}(r' - h') .$$

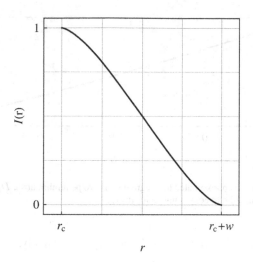

Figure 4.5: Intensity in the penumbra, $I(r)$, according to Eq. (4.6).

over the penumbra area, and the approximated dose $D_{\text{p,app}}$ are compared in the following. The dose obtained by integration of $I(r)$ over the penumbra area, A_p, is

$$D_p = \int I(r)\, A_p\, dA = \int_{r_c}^{r_c+w} I(r)\, 2\pi r\, dr\,, \tag{4.7}$$

while the approximated dose is

$$D_{\text{p,app}} = \frac{1}{2} A_p I_c = \frac{1}{2}\left((r_c + w)^2 - r_c^2\right)\pi\,. \tag{4.8}$$

The ratio of approximated to integrated penumbra dose is shown in Fig. 4.6 as a function of the penumbra width to core radius ratio, w/r_c. For a penumbra width being smaller or equal 11 % of the core radius, the integrated penumbra dose is underestimated by less than 2 %. Within this boundary the approximated penumbra dose, $D_{\text{p,app}}$, is considered a good approximation.

The relative dose to the penumbra, D_p^*, can therefore be written as

$$D_p^* = \frac{D_{\text{p,app}}}{D_c} = \frac{I_c \frac{1}{2} A_p}{I_c A_c} = \frac{1}{2}\frac{\left((r_c + w)^2 - r_c^2\right)}{r_c^2}\,, \tag{4.9}$$

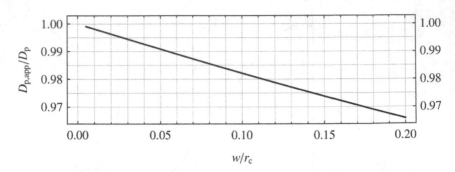

Figure 4.6: The ratio of approximated to integrated relative penumbra dose, $D_{p,app}/D_p$, as a function of the penumbra width to core radius ratio, w/r_c.

where D_c and I_c are the core dose and intensity, respectively.

4.3.2 Equivalent pressure

Generally, the wall collision rate, J_N, in a gas of pressure p is defined as

$$J_N = \frac{n\,\bar{v}}{4} = \frac{p\,\bar{v}}{4k_B T}\,, \quad [J_N] = \text{particles}\,\text{m}^{-2}\text{s}^{-1}\,. \tag{4.10}$$

Therefore, the intensity, I [particles $\text{m}^{-2}\text{s}^{-1}$], of a beam at a certain distance from its source can be translated into an equivalent pressure, denoted by $\tilde{p}$, and given by

$$\tilde{p} = \frac{4k_B T}{\bar{v}} I\,. \tag{4.11}$$

For practical reasons the intensities of the MB will be calculated as equivalent pressures, denoted by the ˜ above the letter p, but still be referred to simply as "pressures".

4.3.3 Molecular Beam intensities and doses

Figure 4.7 shows a schematic of the doses in the various regions of the MB on the sample plane. In the MB core the MB has the MB core pressure, $\tilde{p}_{MBc}$, and the

MB core dose, D_{MBc}, is received. The MB penumbra receives the MB penumbra dose, D_{MBp}, and the PS1 penumbra receives the PS1 penumbra dose, D_{PS1p}. The MB edge dose, D_{edge}, is defined as the sum of the MB penumbra dose and the PS1 penumbra dose,

$$D_{\text{edge}} = D_{\text{MBp}} + D_{\text{PS1p}}. \tag{4.12}$$

Outside of the PS1 penumbra the MB has the MB background pressure, $\tilde{p}_{\text{MBB}}$, caused by the effusing gas from PS2. In the MB penumbra the pressure decreases from $\tilde{p}_{\text{MBc}}$ to the pressure of the PS1 core, $\tilde{p}_{\text{PS1c}}$ (not shown); note that the PS1 core coincides with the MB core. In the PS1 penumbra the pressure decreases from $\tilde{p}_{\text{PS1c}}$ to $\tilde{p}_{\text{MBB}}$.

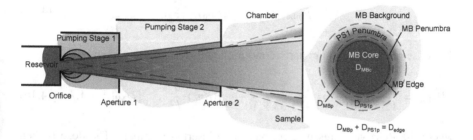

Figure 4.7: Denotation of doses in the different areas of the MB on the sample.

Molecular Beam core pressure and dose

The intensity in the MB core follows from Eq. (3.3), the intensity in forward direction of an orifice, to

$$I = \frac{p_{\text{r}}}{k_{\text{B}}T} \cdot \frac{\overline{v}\,\sigma}{4\pi} \cdot d\Omega = \frac{p_{\text{r}}}{k_{\text{B}}T} \cdot \frac{\overline{v}\,d_{\text{o}}^2}{16} \cdot \frac{1}{l_{\text{os}}^2}, \tag{4.13}$$

where the area of the Orifice, $\sigma = d_o^2 \pi / 4$, and the distance between orifice and sample, l_{os}, were inserted. This intensity is equivalent to a pressure, $\tilde{p}$, given by

$$\tilde{p} = I \frac{4 k_B T}{\bar{v}} = \frac{1}{4} p_r \frac{d_o^2}{l_{os}^2} . \tag{4.14}$$

Equation (4.14) is strictly valid only in the molecular flow limit, $Kn_o \gg 1$. For lower Kn_o, i.e. higher Reservoir pressures, the increasing conductance of the Orifice in the transition flow regime (see Section 2.2.3, Page 15) has to be considered. Therefore Eq. (4.14) is multiplied by $\gamma(Kn_o)$, the ratio of the pressure dependent orifice conductance to the molecular orifice conductance (see Eq. (2.30), Page 16), giving the final expression for the MB core pressure:

$$\tilde{p}_{MBc} = \frac{1}{4} p_r \cdot \gamma(Kn_o) \cdot \frac{d_o^2}{l_{os}^2} . \tag{4.15}$$

The dose rate in Langmuir[2] per second in the MB core is then given by

$$\dot{D}_{MBc} = \frac{\tilde{p}_{MBc}}{1.33 \times 10^{-6}} = \frac{1}{4} \cdot \frac{p_r}{1.33 \times 10^{-6}} \cdot \gamma(Kn_o) \cdot \frac{d_o^2}{l_{os}^2} , \quad [\dot{D}_{MBc}] = \mathrm{L/s} , \tag{4.16}$$

where the Reservoir pressure, p_r, has to be inserted in mbar. The dose is then obtained by multiplying the dose rate with the dosing time:

$$D_{MBc} = \dot{D}_{MBc} \cdot t . \tag{4.17}$$

Molecular Beam penumbra dose

The relative MB penumbra dose is obtained using the general relation between penumbra width and relative penumbra dose (Eq. (4.3), Page 31), giving

$$D_{MBp}^* = \frac{D_{MBp}}{D_{MBc}} = \frac{1}{2} \frac{(w_{MBp} + r_{MBc})^2 - r_{MBc}^2}{r_{MBc}^2} , \tag{4.18}$$

---

[2]One Langmuir is defined as the exposure to gas with a pressure of 1.33×10^{-6} mbar for the duration of one second.

where w_{MBp} is the MB penumbra width.

4.3.4 Geometry

Figure 4.8 shows a geometric model of the Orifice, Aperture 1, Aperture 2 and the sample. Table 4.1 describes the used symbols. In this section general expressions are derived from the geometry, which will be used in the calculation of the MB properties.

Aperture 1 and Aperture 2 diameter

The aperture diameters are defined by Orifice diameter, the required MB core diameter and the position of the apertures.

Aperture 1 A possible miss-alignment of Orifice, Aperture 1 and Aperture 2 due to their respective concentricity tolerances (Δc_{o}, Δc_{a1}, Δc_{a2}) arising from the manufacturing process is accounted for in the Aperture 1 diameter: Orifice and Aperture 2 are assumed to be shifted in the same direction by their respective concentricity tolerance, Aperture 1 is assumed to be shifted in the exact opposite direction by its concentricity tolerance. This worst case miss-alignment is compensated for by a larger Aperture 1 diameter. To derive the Aperture 1 radius one first assumes Aperture 1 to be perfectly concentric giving the preliminary Aperture 1 radius, r'_{a1},

$$\frac{r'_{\mathrm{a1}} - r_{\mathrm{o}} - \Delta c_{\mathrm{o}}}{l_{\mathrm{o\,a1}}} = \frac{r_{\mathrm{a2}} + \Delta c_{\mathrm{a2}} - r'_{\mathrm{a1}}}{l_{\mathrm{a1\,a2}}}$$

$$r'_{\mathrm{a1}} = \frac{l_{\mathrm{o\,a1}}(r_{\mathrm{a2}} + \Delta c_{\mathrm{a2}}) + l_{\mathrm{a1\,a2}}(r_{\mathrm{o}} + \Delta c_{\mathrm{o}})}{l_{\mathrm{o\,a1}} + l_{\mathrm{a1\,a2}}} .$$

Symbol	Description
d_o (r_o)	Orifice diameter (radius)
d_{a1} (r_{a1})	Aperture 1 diameter (radius)
d_{a2} (r_{a2})	Aperture 2 diameter (radius)
$l_{o\,a1}$	Orifice–Aperture 1 distance
$l_{o\,s}$	Orifice–sample distance
$l_{a2\,s}$	Aperture 2–sample distance
r_{MBc} (d_{MBc})	MB core radius (diameter)
w_{MBp}	MB penumbra width
w_{PS1p}	PS1 penumbra width
w_{a2c}	Intercepted core width
w_{a2p}	Intercepted penumbra width
$l_{a1\,a2}$	Aperture 1–Aperture 2 distance

Table 4.1: Notation used in the geometry schematic (Fig. 4.8) and the calculation.

The final expression for r_{a1} is obtained by adding the concentricity tolerance to r'_{a1}, giving

$$r_{a1} = r'_{a1} + \Delta c_{a1} = \frac{l_{o\,a1}(r_{a2} + \Delta c_{a2}) + l_{a1\,a2}(r_o + \Delta c_o)}{l_{o\,a1} + l_{a1\,a2}} + \Delta c_{a1}\,. \qquad (4.19)$$

Aperture 2   For the Aperture 2 radius one finds

$$\frac{r_{MBc} - r_o}{l_{o\,s}} = \frac{r_{a2} - r_o}{l_{o\,s} - l_{a2\,s}}$$

$$r_{a2} = (r_{MBc} - r_o)\frac{l_{o\,s} - l_{a2\,s}}{l_{o\,s}} + r_o\,. \qquad (4.20)$$

MB penumbra width

w_{MBp} is defined by the Orifice diameter and the distances between Orifice, Aperture 2 and the sample by using Eq. (4.3), Page 31, giving

$$w_{MBp} = d_o\frac{l_{a2\,s}}{l_{o\,a2}}\,. \qquad (4.21)$$

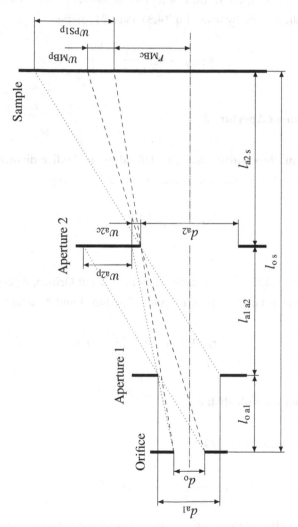

Figure 4.8: Geometry schematic of the MB. For a description of the symbols see Table 4.1. Lengths, diameters, etc. are not to scale.

PS1 penumbra width

w_{PS1p} is defined by the Aperture 1 diameter and the distances between Aperture 1, Aperture 2 and the sample by using Eq. (4.3), Page 31, giving

$$w_{PS1p} = d_{a1} \frac{l_{a2\,s}}{l_{a1\,a2}}.$$ (4.22)

Intercepted beam at Aperture 2

Intercepted penumbra width w_{ap2p} is defined by the Orifice diameter and the distances between Orifice, Aperture 1 and Aperture 2 by using Eq. (4.3), Page 31, giving

$$w_{a2p} = d_o \frac{l_{a1\,a2}}{l_{o\,a1}}.$$ (4.23)

Intercepted core width w_{ap2c} is defined by the radii of Orifice, Aperture 1, and Aperture 2 and the distances between Orifice, Aperture 1 and Aperture 2, giving

$$w_{a2c} = r_{a1} + (r_{a1} - r_o) \frac{l_{a1\,a2}}{l_{o\,a1}} - r_{a2}.$$ (4.24)

Optimal Orifice to sample distance

The MB core pressure, given by Eq. (4.15), is

$$\tilde{p}_{MBc} = \frac{1}{4} p_r \cdot \gamma(Kn_o) \cdot \frac{d_o^2}{l_{o\,s}^2}.$$ (4.25)

To maximize the MB core pressure while keeping the two conditions in the MB requirements, (1) $Kn_o \geq 2.5$ and (2) $D_{edge}^* \leq 2.5\,\%$, p_{MBc} needs to be expressed by Kn_o and D_{edge}^*. Therefore, first the Reservoir pressure is related to Kn_o and d_o, then d_o is related to D_{edge}^*.

Considering the first condition, Kn_o is expressed by the Reservoir pressure according to

$$\lambda_r = Kn_o \cdot d_o \quad \text{and} \quad p_r = \frac{\eta \pi \bar{v}}{4} \frac{1}{\lambda_r}, \tag{4.26}$$

giving

$$p_r = \frac{\eta \pi \bar{v}}{4} \frac{1}{Kn_o \cdot d_o}. \tag{4.27}$$

Inserting Eq. (4.27) into Eq. (4.25) gives

$$\tilde{p}_{\mathrm{MBc}} = \frac{\eta \pi \bar{v}}{16} \frac{1}{Kn_o \cdot d_o} \cdot \gamma(Kn_o) \cdot \frac{d_o^2}{l_{o\,s}^2} = \frac{\eta \pi \bar{v}}{16} \frac{1}{Kn_o} \cdot \gamma(Kn_o) \cdot \frac{d_o}{l_{o\,s}^2}. \tag{4.28}$$

For the second condition, $D_{\mathrm{edge}}^* = D_{\mathrm{MBp}}^* + D_{\mathrm{PS1p}}^*$ is approximated by D_{MBp}^*, assuming that $D_{\mathrm{MBp}}^* \gg D_{\mathrm{PS1p}}^*$. The MB penumbra width is expressed by D_{MBp}^*, Eq. (4.18), giving

$$w_{\mathrm{MBp}} = r_{\mathrm{MBc}} \left(\sqrt{1 + 2D_{\mathrm{MBp}}^*} - 1 \right). \tag{4.29}$$

Inserting w_{MBp} from Eq. (4.29) into Eq. (4.21), Page 38, yields the following relation of the Orifice diameter to D_{MBp}^*:

$$d_o = \frac{l_{o\,s} - l_{a2\,s}}{l_{a2\,s}} r_{\mathrm{MBc}} \left(\sqrt{1 + 2D_{\mathrm{MBp}}^*} - 1 \right), \tag{4.30}$$

where $l_{o\,a2} = l_{o\,s} - l_{a2\,s}$ was used.

Inserting d_o from Eq. (4.30) into Eq. (4.28) finally yields

$$\tilde{p}_{\mathrm{MBc}} = \frac{\eta \pi \bar{v}}{16} \frac{1}{Kn_o} \cdot \gamma(Kn_o) \cdot r_{\mathrm{MBc}} \left(\sqrt{1 + 2D_{\mathrm{MBp}}^*} - 1 \right) \frac{l_{o\,s} - l_{a2\,s}}{l_{a2\,s}\, l_{o\,s}^2}. \tag{4.31}$$

When treating Kn_o, D_{MBp}^* and r_{MBc} as constants, p_{MBc} is only dependent on the distances between Orifice and Aperture 2, and Orifice and sample. $l_{a2\,s}$ should obviously be as small as possible and will be chosen according to the constructional possibilities. This leaves $l_{o\,s}$ as only parameter to be optimized. Therefore the derivative of p_{MBc} with respect to l_{os} is formed and set equal to zero, yielding

$$\frac{\partial}{\partial l_{os}} \tilde{p}_{MBc} \bigg|_{l_{a2s},Kn_o,D^*_{MBp},r_{MBc}} = 0 \Rightarrow l_{os} = 2\, l_{a2s}\,, \qquad (4.32)$$

revealing that the distance from Orifice to the sample should be twice the distance from Aperture 2 to the sample in order to maximize $\tilde{p}_{MBc}$ while keeping the conditions for Kn_o and D^*_{MBp}. Figure 4.9 shows the dependence of $\tilde{p}_{MBc}$ on l_{a2s} and l_{os}.

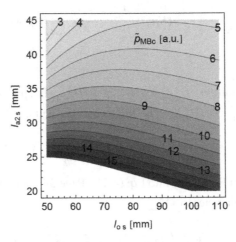

Figure 4.9: Contour plot of the MB core pressure, $\tilde{p}_{MBc}$, as a function of Aperture 2 to sample distance, l_{a2s}, and Orifice to sample distance, l_{os}, according to Eq. (4.31).

4.3.5 Molecular Beam dimensions

Orifice to sample distance

For the MB it is required that $l_{a2s} > 30\,\text{mm}$, which is the very limit for this dimension. However, for manufacturing reasons the distance between Aperture 2 and sample was set to $l_{a2s} = 39\,\text{mm}$. Therefore the ideal value l_{os} according to Eq. (4.32) would be $2l_{a2s} = 78\,\text{mm}$, but was chosen to be 90 mm to have more space for pumping and the components of the shutter.

Orifice diameter

With $d_{MBc} = 3.5\,$mm, $D^*_{edge} = 2.5\,\%$, $Kn_o = 2.5$, $l_{a2\,s} = 39\,$mm and $l_{o\,s} = 90\,$mm the upper limit for the Orifice diameter follows from Eq. (4.30) to be 57 μm. However, an Orifice diameter of $d_o = 50\,$μm was chosen since that was the next commercially available orifice size.

Aperture 1 distance

The distance from the Orifice to Aperture 1 was set to $l_{o\,al} = 14.4\,$mm. This distance is big enough to allow for reasonable pumping of the volume between the Orifice and Aperture 1, and small enough to keep the diameter of Aperture 1 small in order to keep the conductance between the pumping stages small.

Concentricity tolerances

The concentricity tolerances $\Delta c_o = 50\,$μm and $\Delta c_{a1} = \Delta c_{a2} = 10\,$μm were assumed.

Summary

The remaining dimensions follow from the equations derived in the previous section and are shown in Table 4.2.

4.3.6 Pressures in the pumping stages and the chamber

The base pressures in the pumping stages and the chamber (due to outgassing of the walls, leak rate of the seals) are not known; therefore, only expressions for the pressure increases caused by the MB can be derived.

Quantity	Value	Quantity	Value	Equation
d_o	50 μm	d_{a1}	0.7 mm	(4.19)
d_{MBc}	3.5 mm	d_{a2}	2.0 mm	(4.20)
$l_{o\,a1}$	14.4 mm	w_{MBp}	38.2 μm	(4.21)
$l_{o\,a2}$	51.0 mm	w_{PS1p}	0.75 mm	(4.22)
$l_{o\,s}$	90.0 mm	w_{a2c}	0.18 mm	(4.23)
$l_{a1\,s}$	75.6 mm	w_{a2p}	0.13 mm	(4.24)
$l_{a1\,a2}$	36.6 mm			
$l_{a2\,s}$	39.0 mm			

Table 4.2: Dimensions of the MB.

The pressure increase in the pumping stages is defined by the gas loads originating from the previous stages (Reservoir and PS1, respectively) and the effective pumping speeds in the pumping stages.

Pumping Stage 1 The main gas load in PS1 consists of gas effusing from the Reservoir through the Orifice. Although the Orifice was treated as an ideal orifice with $l = 0$ for the calculation of the MB core pressure, the finite thickness of $l_o = 20\,\mu m$ (see Section 4.4.2, Page 54) has to be considered for the calculation of the gas throughput. Therefore the conductance of the ideal Orifice, $C_{O,o}$, is multiplied with the transmission probability for an equivalent tube $P_T(l_o, d_o)$ according to Eq. (2.8), Page 7. The small part of the gas load which leaves through Aperture 1 is neglected. Therefore the pressure increase in PS1 is given by

$$p_{1i} = \frac{\gamma(Kn_o) C_{O,o} P_T(l_o, d_o) p_r}{S_{PS1}}, \tag{4.33}$$

where $\gamma(Kn_o)$ is the ratio of the pressure dependent orifice conductance to the molecular orifice conductance (see Eq. (2.30), Page 16), and S_{PS1} the effective pumping speed in PS1.

Pumping Stage 2 The gas load in PS2 consists of the gas effusing from PS1 through Aperture 1 and the gas in the molecular beam formed by the Orifice and

Aperture 1, that is intercepted at Aperture 2 and scattered into PS2.

On the wall of Aperture 2, which is facing PS2, two areas are distinguished: the core ring of width w_{a2c}, which intercepts a portion of the molecular beam core, and the penumbra ring of width w_{a2p}, which intercepts the penumbra, see Fig. 4.10.

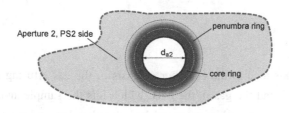

Figure 4.10: Intercepted molecular beam areas at the PS2 side of Aperture 2. Figure not to scale.

The gas load from the core ring is given by the pressure in the molecular beam multiplied by the orifice conductance of the core ring area:

$$q_{cr} = \tilde{p}_{cr} C_{0,A_{cr}} = p_r \frac{r_o^2}{l_{o\,a2}^2} C_{0,A_{cr}}, \tag{4.34}$$

where Eq. (3.3), Page 19, for the forward intensity of an effusive source was used. The area of the core ring is

$$A_{cr} = \left((r_{a2} + w_{a2c})^2 - r_{a2}^2 \right) \pi. \tag{4.35}$$

By analogy to the penumbra dose (see Section 4.3.1), one can also approximate the gas load from the penumbra by assuming the full core pressure on half the penumbra area. Therefore the gas load from the penumbra ring is given by

$$q_{pr} = p_r \frac{r_o^2}{l_{o\,a2}^2} \frac{1}{2} C_{0,A_{pr}}. \tag{4.36}$$

where the area of the penumbra ring is

$$A_{pr} = \left[(r_{a2} + w_{a2c} + w_{a2p})^2 - (r_{a2} + w_{a2c})^2 \right] \pi. \tag{4.37}$$

Therefore the pressure increase in PS2, p_{2i}, is given by

$$p_{2i} = \frac{1}{S_{PS2}} \left(C_{a1} p_{1i} + p_r \frac{r_o^2}{l_{o\,a2}^2} (C_{O,A_{cr}} + \frac{1}{2} C_{O,A_{pr}}) \right), \tag{4.38}$$

where C_{a1} is the conductance of Aperture 1.

Chamber The gas load to the chamber consists of the gas effusing from PS2 through Aperture 2 and the gas in the MB, which hits the sample and is subsequently either contributing when scattered into the chamber or not contributing when frozen on the sample, depending on experimental conditions (sample temperature, sticking probability); therefore, the pressure increase in the chamber in the freezing case, $p_{ci,freeze}$, is given by

$$p_{ci,freeze} = \frac{1}{S_c} (C_{a2} p_{2i}) \tag{4.39}$$

and the pressure increase in the scattering case, p_{ci}, is given by

$$p_{ci} = \frac{1}{S_c} \left(C_{a2} p_{2i} + C_{O,d_{MBc}} \tilde{p}_{MBc} \right) \tag{4.40}$$

where C_{a2} is the conductance of Aperture 2 and $C_{O,d_{MBc}}$ the orifice conductance of the MB core area.

4.3.7 Effusing gas from the pumping stages

Each pumping stage acts as an effusive beam source on its own. While the effusing gas from PS1 hits the sample surface only due to Aperture 2, gas from PS2 effuses unhindered to all surfaces in the chamber.

Pumping Stage 1 The PS1 core pressure is given by

$$\tilde{p}_{\mathrm{PS1c}} = p_{1\mathrm{i}} \frac{r_{\mathrm{a1}}^2}{l_{\mathrm{a1\,s}}^2}, \tag{4.41}$$

where Eq. (3.3), Page 19, for the forward intensity of an effusive source was used. The ratio of the PS1 core to the MB core pressure is given by

$$E_{\mathrm{PS1}} = \frac{\tilde{p}_{\mathrm{PS1c}}}{\tilde{p}_{\mathrm{MBc}}} = \frac{p_{1\mathrm{i}}}{p_{\mathrm{r}}} \frac{r_{\mathrm{a1}}^2}{r_{\mathrm{o}}^2} \frac{l_{\mathrm{o\,s}}^2}{l_{\mathrm{a1\,s}}^2}. \tag{4.42}$$

The relative dose to the PS1 penumbra, D_{PS1p}^*, is therefore

$$D_{\mathrm{PS1p}}^* = \frac{D_{\mathrm{PS1p}}}{D_{\mathrm{MBc}}} = E_{\mathrm{PS1}} \frac{1}{2} \frac{(r_{\mathrm{MBc}} + w_{\mathrm{PS1p}})^2 - r_{\mathrm{MBc}}^2}{r_{\mathrm{MBc}}^2}, \tag{4.43}$$

where it was assumed that the PS1 core radius on the sample is equal to r_{MBc} and the approximation for the penumbra dose was used, see Section 4.3.1, Page 30.

Pumping Stage 2 The pressure of gas effusing from PS2 in the forward direction at the sample is by definition the background pressure of the MB in the sample plane, given by

$$\tilde{p}_{\mathrm{MBB}} = p_{2\mathrm{i}} \frac{r_{\mathrm{a2}}^2}{l_{\mathrm{a2\,s}}^2}, \tag{4.44}$$

where Eq. (3.3), Page 19, for the forward intensity of an effusive source was used. The ratio of the background pressure to the MB core pressure is given by

$$E_{\mathrm{MBB}} = \frac{\tilde{p}_{\mathrm{MBB}}}{\tilde{p}_{\mathrm{MBc}}} = \frac{p_{2\mathrm{i}}}{p_{\mathrm{r}}} \frac{r_{\mathrm{a2}}^2}{r_{\mathrm{o}}^2} \frac{l_{\mathrm{o\,s}}^2}{l_{\mathrm{a2\,s}}^2}. \tag{4.45}$$

4.3.8 Molecular Beam properties

The Reservoir gas is assumed to be N_2 at $T = 300\,\mathrm{K}$ for the calculation of the MB properties. Furthermore the knowledge of the effective pumping speeds in PS1 and PS2, S_{PS1} and S_{PS2}, and the conductances of Aperture 1 and Aperture 2, C_{a1} and C_{a2}, are necessary. The effective pumping speeds of the pumping stages and said conductances are estimated to be $S_{PS1} = 24.2\,l/s$, $S_{PS2} = 27.6\,l/s$, $C_{a1} = 18.6\,cm^3/s$ and $C_{a2} = 171.2\,cm^3/s$ (Section 4.4.6, Page 68).

In Table 4.3 the relative doses of the MB and the relative MB background pressure are shown. The relative dose to the MB edge, $D^*_{edge} = 2.31\,\%$, is smaller than the required maximum of 2.5 % due to the chosen Orifice diameter of $50\,\mu m$, which is smaller than the calculated diameter of $57\,\mu m$. The assumption that $D^*_{MBp} \gg D^*_{PS1p}$ was correct. The ratio of the MB background pressure to the MB core pressure is 0.006 %, which means that the MB core pressure is 4 orders of magnitude higher than the background pressure.

Quantity	Value	Equation
D^*_{edge}	2.31 %	(4.12)
D^*_{MBp}	2.21 %	(4.18)
D^*_{PS1}	0.10 %	(4.43)
E_{MBB}	0.006 %	(4.45)

Table 4.3: Relative doses of the MB and the relative MB background pressure.

In the following, the Reservoir pressure dependent properties of the MB are calculated.

Standard Reservoir pressure

The standard Reservoir pressure, which corresponds to a Knudsen number at the orifice of $Kn_o = 2.5$ is given by

$$p_r = \frac{1}{2.5} \frac{1}{d_o} \frac{\eta \bar{v} \pi}{4} = 5.4 \times 10^{-1}\,mbar\,, \tag{4.46}$$

where Eq. (4.27) was used. Table 4.4 shows the MB properties for the standard Reservoir pressure. The MB core pressure is equivalent to a dosing rate of approximately 30 seconds per Langmuir, which is quite low but acceptable.

Quantity	Value		Equation
MB dose rate and pressures			
$\dot{D}_{MBc}$	0.032	L/s	(4.16)
$\tilde{p}_{MBc}$	4.3×10^{-8}	mbar	(4.15)
$\tilde{p}_{PS1c}$	8.2×10^{-11}	mbar	(4.41)
$\tilde{p}_{MBB}$	2.3×10^{-12}	mbar	(4.44)
Pressure increases			
p_{1i}	3.8×10^{-6}	mbar	(4.33)
p_{2i}	3.5×10^{-9}	mbar	(4.38)
$p_{ci,freeze}$	2.0×10^{-12}	mbar	(4.39)
p_{ci}	1.7×10^{-10}	mbar	(4.40)

Table 4.4: MB properties for the standard Reservoir pressure $p_r = 5.4 \times 10^{-1}$ mbar.

Elevated Reservoir pressure

Table 4.5 shows the MB properties when the standard Reservoir pressure is increased 100 times to $p_r = 54.0$ mbar. At that pressure $Kn_o = 0.025$, so the Orifice is already in the transition flow regime. Therefore the MB penumbra width may be larger than the theoretical value due to a virtual larger Orifice area. The dose rate $\dot{D}_{MBc} = 4.7$ L/s is equivalent to a dosing time of 0.21 s per Langmuir, and is close to the required maximum dose rate of 5 L/s.

Quantity	Value		Equation
MB dose rate and pressures			
$\dot{D}_{MBc}$	4.7	L/s	(4.16)
$\tilde{p}_{MBc}$	6.3×10^{-6}	mbar	(4.15)
$\tilde{p}_{PS1c}$	1.2×10^{-8}	mbar	(4.41)
$\tilde{p}_{MBB}$	3.4×10^{-10}	mbar	(4.44)
Pressure increases			
p_{1i}	5.7×10^{-4}	mbar	(4.33)
p_{2i}	5.2×10^{-7}	mbar	(4.38)
$p_{ci,freeze}$	3.0×10^{-10}	mbar	(4.39)
p_{ci}	2.5×10^{-8}	mbar	(4.40)

Table 4.5: MB properties for the Reservoir pressure $p_r = 54.0$ mbar.

Apart from the changing properties of the MB due to the transition to the viscous flow regime, the highest Reservoir pressure is also limited by the material strength of the thin-walled Orifice (see Section 4.4.2, Page 54). It is estimated that a Reservoir pressure of 200 mbar should not be exceeded.

4.3.9 Capillary pressure

Gas is leaked into the Reservoir through capillaries connected to the Piezo leak vales. The Reservoir is pumped by one of the pumping orifices (PO). The pressure at the entrance of the capillary, p_{cap} (capillary pressure), required to establish a certain Reservoir pressure, p_r, is obtained by solving the equation of the gas flows into and out of the Reservoir for p_{cap}:

$$\underbrace{(p_{cap} - p_r) \cdot C_T(p_{cap}, p_r)}_{\text{Flow in}} =$$

$$p_r \cdot \underbrace{\left[\underbrace{\gamma(Kn_o)C_{O,o} P_T(t_o, d_o)}_{\text{MB}} + \underbrace{\gamma(Kn_{PO})C_{O,PO}}_{\text{PO}} \right]}_{\text{Flow out}}, \quad (4.47)$$

where $C_T(p_{cap}, p_r)$ is the conductance of the capillary according to Eq. (2.35), Page 17; the term "MB" is the conductance of the Orifice (see Section 4.3.6, Page 43); the term "PO" is the conductance of the PO with diameter d_{PO}; Kn_{PO} is the Knudsen number at the PO.

Figure 4.11 shows p_{cap} for a capillary with diameter $d_{cap} = 0.508$ mm and length $l_{cap} = 1.2$ m (see Reservoir dimension in Section 4.4.5, Page 67) and $d_{PO} = 800$ μm. A capillary pressure of roughly 10 to 200 mbar spans a Reservoir pressure of 10^{-2} to 2 mbar.

Figure 4.12 shows the p_{cap}, for the same capillary dimensions but for $d_{PO} = 150$ μm. A capillary pressure of roughly 10 to 200 mbar spans a Reservoir pressure of 3×10^{-1} to 50 mbar.

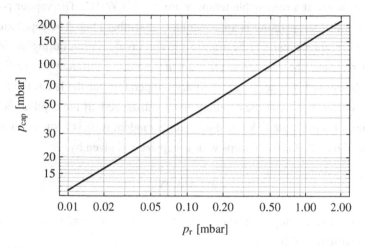

Figure 4.11: Required capillary pressure, p_{cap}, as a function of the Reservoir pressure, p_r, for a pumping orifice diameter of $d_{PO} = 800\,\mu m$, according to Eq. (4.47).

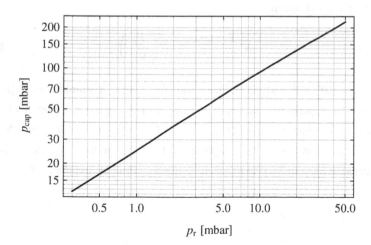

Figure 4.12: Required capillary pressure, p_{cap}, as a function of the Reservoir pressure, p_r, for pumping orifice diameter of $d_{PO} = 150\,\mu m$, according to Eq. (4.47).

Water   In any case the necessary maximum capillary pressure of 200 mbar allows the use of water at a reasonable temperature of $T = 60\,°C$. The vapour pressure of water at that temperature is approximately 200 mbar [17]. The capillaries will be heated by direct resistive heating to prevent condensation during dosing and to properly clean the capillaries after dosing, before a different gas or liquid is dosed. To estimate the necessary capillary temperature for the latter, the time for a water molecule to leave the capillary from a distance half the capillary's length from the exit is calculated. A one-dimensional random walk is assumed, hence the distance X travelled after N steps with a stepsize s is given by

$$X^2 \approx s^2 N . \tag{4.48}$$

If the time between steps is τ, then $N = t/\tau$, and the average time, t, it takes to travel the distance X is

$$t \approx \frac{X^2}{s^2}\tau . \tag{4.49}$$

The time between steps is the average residence time, τ_r, of a water molecule on the capillary wall, which is equal to the reciprocal rate constant for desorption, k_d (see Eq. (3.1), Page 19), given by

$$\tau_r = \frac{1}{k_d} = A^{-1}\exp\left(\frac{E_d}{RT}\right) , \tag{4.50}$$

where $A = 10^{13}\,s^{-1}$ and $E_d = 96\,kJ/mol$ for water on stainless steel [20] are assumed. The stepsize is assumed to be one capillary diameter. Figure 4.13 shows the time, t, according to Eq. (4.49) for $X = \frac{1}{2}l_{cap} = 0.6\,m$, $s = d_{cap} = 0.508\,mm$ and $\tau = \tau_r$ as a function of the temperature. One can see that the capillaries have to be heated at least to $300\,°C$ to reach an average time of less than $100\,s$.

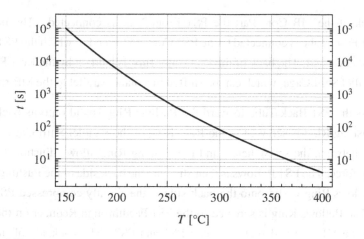

Figure 4.13: Average time for a water molecule to leave a capillary according to a random walk model. Shown is the average time, t, according to Eq. (4.49) for $X = \frac{1}{2}l_{cap} = 0.6\,m$, $s = d_{cap} = 0.508\,mm$ and $\tau = \tau_r$ as a function of the temperature.

4.4 Construction

First, a brief overview of the MB construction is given, then the components are described in the following sections.

4.4.1 Overview

Figure 4.14 shows a 3D model of the Molecular Beam. Components of the Reservoir (green), PS1 (red) and PS2 (blue) are colored accordingly.

The Orifice Inset, Aperture 1, Aperture 2 and the Shutter Assembly are held by the MB Core Part. The front of the MB Core Part is pressed against the PS2 Shell, which consists of the PS2 Tube leading from the MB Reducer into the chamber and the PS2 Cone, closing the PS2 Tube and leaving only an opening for Aperture 2 to protrude into the chamber. The MB Tee together with the PS2 Shell completes PS2.

Inside the PS2 Shell and the MB Tee the tubing of PS1 is accommodated. To the back of the MB Core Part the PS1 Front Tube is connected. The back of the PS1 Front Tube is connected to the PS1 Cone, to the end of which the PS1 Back Tube is connected. The back of the PS1 Back Tube is held on by the Rear Ring of the Parallel Shift Cage, which can be shifted in a plane parallel to the MB reducer.

To press the PS1 Back Tube forward the Pushing Ring is slid onto its back side. The Pushing Ring is screwed to the Rear Ring of the Parallel Shift Cage, where springs between the screw heads and the Pushing Ring allow adjustment of the pushing force. The PS1 Bellows is then slid onto the back side of the Pushing Ring. The Bellows Ring is slid onto the back side of the slightly compressed PS1 Bellows. The Bellows Ring is screwed to the PS1 Feedthrough Reducer on the back side of the MB Tee and seals between PS1 and PS2. The backside of the PS1 Feedthrough Reducer is connected to the MB Cross which completes PS1.

The inside section of the Reservoir consists of the Orifice Inset, the Reservoir Tube, the Reservoir Bellows, the Bellows Adapter and the Reservoir Feedthrough Reducer. The Reservoir Tube and the Reservoir Bellows are located inside of PS1 and connect the Orifice Inset to the Reservoir Feedthrough Reducer at the back flange of the MB Cross. A standard CF 40 to CF 16 zero length reducer is screwed to the backside of the Reservoir Feedthrough Reducer, providing the access point for the outside section of the Reservoir.

4.4.2 Molecular Beam Core Part

The MB Core Part holds the Orifice Inset, Aperture 1, Aperture 2 and the Shutter Assembly. Its overall length and diameter are 46.9 mm and 54 mm, respectively. The MB Core Part, the Orifice Inset, Aperture 1, Aperture 2 and the Ball Bearing Envelope (a component of the Shutter Assembly) were machined by a company[3] from very low magnetic stainless steel (1.4429-ESU/316LN-ESR). Figure 4.15 shows the MB Core Part. The side in the front of the MB Core Part has 5 openings for pumping the volume behind Aperture 2 and one opening for the Shutter.

---

[3]Gerhard Rauch GmbH, Gewerbepark 1, A-3452 Trasdorf

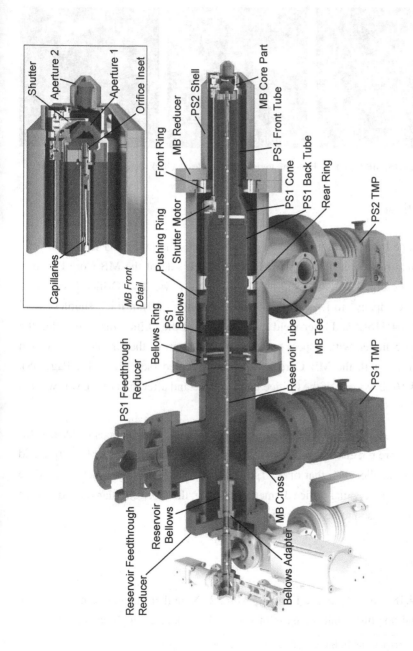

Figure 4.14: 3D model of the Molecular Beam. Components of the Reservoir (green), Pumping Stage 1 (red) and Pumping Stage 2 (blue) are colored accordingly.

Figure 4.15: MB Core Part. From left to right: front, side, and back view of the MB Core Part.

Orifice Inset

The Orifice Inset is a tube-shaped part which holds the Orifice. The Orifice Inset is mounted by sliding it into place from the back side of the MB Core Part into its seat (see Fig. 4.15) and securing it with three srews. Two Orifice Insets were sent to a company[4] to laser drill a 50 µm hole into a 20 µm thick stainless steel foil (1.4404/316L) and laser-weld it to the front of the Orifice Inset such that the Orifice is centered with respect to the 8 mm wide part of the Orifice Inset, which is in contact with the MB Core Part, when mounted (see Fig. 4.19, Page 58). Figure 4.16 shows one of the Orifice Insets before and after the Orifice was welded to it.

Figure 4.17 shows SEM images of the two Orifices. The Orifices of Orifice Inset 1 and 2 are not circular and their respective equivalent diameter of 38.0 µm and 46.8 µm is smaller than that of a 50 µm hole. Although Orifice Inset 2 would be closer to the conceptual orifice diameter of 50 µm, the more circular shaped Orifice Inset 1 was chosen to be used for testing the MB.

Apertures

Figure 4.18 shows Aperture 1 and Aperture 2. Note that the visible holes are not the actual apertures, but the exits of short tubes added after the actual apertures

---

[4]RJ Lasertechnik GmbH, Boschstraße 20, D-52531 Übach-Palenberg

Figure 4.16: Orifice Inset without and with laser-welded Orifice. Left: Orifice Inset before the Orifice was laser-welded to the front. Right: front part of the Orifice Inset after welding. a) 200 μm thick cover foil for the welding process, b) 20 μm thick foil with laser-drilled orifice in its center.

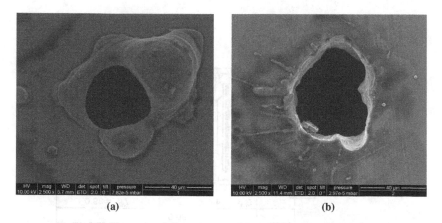

Figure 4.17: SEM images of the Orifices in the Orifice Insets. (a) Orifice of Orifice Inset 1, estimated area of 1136 μm², equivalent to a circular orifice diameter of 38.0 μm. (b) Orifice of Orifice Inset 2, estimated area of 1717 μm², equivalent to a circular orifice diameter of 46.8 μm.

to reduce the conductance (see Section 4.4.7, Page 72 for exact dimensions). The apertures are mounted on the MB Core Part by sliding them into their respective seats (see Fig. 4.19) and securing each with 3 threaded rods.

Figure 4.18: Aperture 1 and Aperture 2. Left: Aperture 1. Right: Aperture 2. Note that the visible holes are not the actual apertures, but the exits of short tubes added after the actual apertures to reduce the conductance. The scale bar applies to both images.

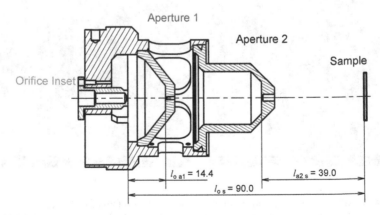

Figure 4.19: Cross section of the MB Core Part with mounted Orifice Inset, Aperture 1 and Aperture 2. Distances in mm.

Shutter

Figure 4.20 shows the Shutter Assembly. The Shutter itself is machined from aluminium (EN AW 6082) and fixed to the Shutter Shaft by 2 screws. The Shutter Shaft is held by the front and rear ball bearing in the Ball Bearing Envelope. The Shutter Shaft can slide freely inside the front and rear ball bearing to compensate differences in thermal expansion during bakeout. A small spring between the front ball bearing and the wide section of the Shutter Shaft pushes the Shutter Shaft against the rear ball bearing into its default position. The Shutter Assembly is mounted by screwing the Ball Bearing Envelope to the MB Core Part.

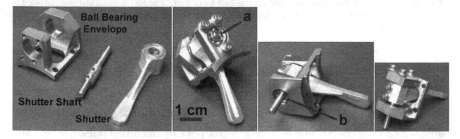

Figure 4.20: Shutter Assembly. a) front ball bearing, b) rear ball bearing. The spring between Shutter Shaft and front ball bearing is not visible.

Figure 4.21 shows the Shutter Assembly mounted on the MB Core Part. The Shutter closes the exit of Aperture 1, when closed. The default distance between the Shutter and Aperture 1 is set by placing an aluminium foil of appropriate thickness between the backside of the Ball Bearing Envelope and the MB Core Part. The default distance of the Shutter to Aperture 1 in the final assembly was estimated to be 20 to 50 μm. The Shutter Shaft is connected to the Motor Shaft by the Shaft Coupling.

4.4.3 Shutter Motor

The shutter is actuated by an electromagnetic motor, which sits in the Motor Block on the PS1 Cone, see Fig. 4.23. A samarium–cobalt magnet (maximum operating

Figure 4.21: Shutter Assembly mounted on the MB Core Part. The PS1 Front Tube is connected to the back of the MB Core Part. Left: Shutter, Shaft Coupling and Aperture 1. Right: Aperture 2 O-ring and Motor Shaft.

temperature $300\,°C$) with a diameter and height of 2 mm (see Fig. 4.22) is fixed to the Motor Lever, which can move between the pole ends of two coils. The Motor Lever has a down position (Shutter open) and an up position (Shutter closed). The Motor Lever changes position by changing the polarity of current pulses, which are applied to the coils. The Motor Lever is fixed to the Motor Shaft, which transfers the movement to the Shutter at the MB Core Part. The Motor Shaft is held in the back of the Motor Block by a ball bearing. The Motor Lever and the Motor Block are made of aluminium.

The coil core is made of ARMCO® iron (Fe > 99.85 %) and has a quadratic cross section with a side length of 2 mm. Around the coil core a double layer of Kapton® insulated copper wire (conductor diameter 0.6 mm) is wound. The edges of the coil cores were filed to protect the insulation of the wire. The coil core ends are machined to a round shape to fit into the respective holes in the Motor Block. The latter is especially important for the inside coil, as its core ends seal between PS1 and PS2. Each coil is secured in place by a small copper plate, which is screwed to a threaded rod. A coil current of 1.0 to 1.6 A is necessary in atmosphere to toggle the position of the Shutter.

To limit the movement of the Motor Lever to the two intended positions and to

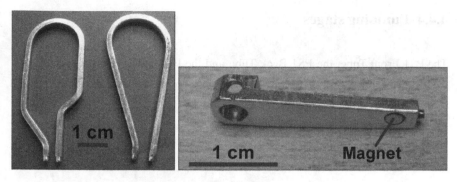

Figure 4.22: Coil cores and Motor Lever of the Shutter Motor. Left: Shutter coils. Right: Motor Lever and magnet, held by a screw.

dampen the oscillations after the shutter was toggled, two Viton pieces are used. The down position is limited by a small Viton piece, which is glued to the Motor Block underneath the Motor Lever. The up position is limited by a small Viton piece, which rests on the Motor Block above the Motor Lever and is held by an aluminium clamp, see Fig. 4.23.

Figure 4.23: Shutter Motor. The ends of the outside and inside coil are located exactly opposite to each other.

4.4.4 Pumping stages

The PS1 Front Tube, the PS1 Back Tube and the PS2 Tube were machined from stainless steel tubes[5] (1.4301/304L) and are electropolished[6] on the in- and outside.

Pumping Stage 2 Shell

Figure 4.24 shows the PS2 Shell and the Rear Ring of the Parallel Shift Cage. The PS2 Tube is welded to the MB Reducer, the Parallel Shift Cage is screwed to the MB Reducer.

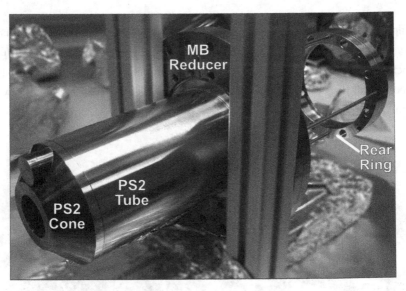

Figure 4.24: PS2 Shell and Rear Ring of the Parallel Shift Cage.

---

[5] outer diameter 54 mm, wall thickness 2 mm (PS1 Front Tube)
   outer diameter 101.6 mm, wall thickness 3 mm (PS1 Back Tube, PS2 Tube)
   Karl Mertl Handelsges.m.b.H., Hähergasse 14, A-2320 Schwechat-Rannersdorf
[6] Henkel Beiz- und Elektropoliertechnik Betriebs GmbH, Stoissmühle 2, A-3830 Waidhofen a.d. Thaya

Pumping Stage 1 Front Assembly

Figure 4.25a shows the PS1 Front Assembly consisting of the MB Core Part, the PS1 Front Tube, the PS1 Cone and the PS1 Back Tube. The PS1 Front Assembly is inserted into the Parallel Shift Cage and the PS2 Shell, see Fig. 4.25b. The Viton O-ring at the front of the MB Core Part (see Fig. 4.21) is pressed between Aperture 2 and the inside of the PS2 Cone and seals between PS2 and the chamber. The position of the PS1 Front Assembly can be adjusted by the four adjustment screws at the Front Ring of the Parallel Shift Cage, pressing on the PS1 Cone, see Fig. 4.26a.

A tungsten filament is placed in PS1 next to the Orifice Inset to shine light through Aperture 1 and Aperture 2. The light spot allows adjustment of the sample position with respect to the MB and provides a visual indication whether the Shutter is opened or closed. The filament can be slightly seen in Fig. 4.31, Page 69. The filament wires are run inside a ceramic tube, which is fixed to the PS1 Cone by the Filament Holder.

Pumping Stage 1 Rear Assembly

The PS1 Rear Assembly consists of the Pushing Ring, the PS1 Bellows and the Bellows Ring. The Pushing Ring is slid onto the end of the PS1 Back Tube. Four screws push the Pushing Ring via springs against the Rear Ring of the Parallel Shift Cage, see Fig. 4.26b. The PS1 Bellows is slid onto the backside of the Pushing Ring and the Bellows Ring is then slid onto the rear end of the PS1 Bellows, see Fig. 4.27.

Figure 4.28 shows the MB Tee, which is screwed to the MB Reducer and encases the PS1 Rear Assembly. The PS1 Feedthrough Reducer is then screwed to the rear flange of the MB Tee. The Bellows Ring is screwed to the MB Reducer by custom-made plate nuts. It seals between PS1 and PS2 by a metal to metal contact.

(a) PS1 Front Assembly.

(b) PS1 Front Assembly inserted into the Parallel Shift Cage.

Figure 4.25: PS1

Figure 4.26: (a) Position adjustment of the MB, showing two of the four adjustment screws (the other two are located on the other side). (b) Pushing Ring at the back of the PS1 Back Tube. Four screws push the Pushing Ring via springs against the Rear Ring of the Parallel Shift Cage.

Figure 4.27: PS1 Rear Assembly.

Wiring

The MB Cross is connected to the backside of the PS1 Feedthrough Reducer. The wires of the shutter and the filament in PS1 are connected to a ceramic D-sub Plug, which lies inside the PS1 Bellows, see Fig. 4.29a. A D-sub cable (plugs of the cable also made of ceramic) connects the D-sub Plug to the D-sub feedthrough at the CF 40 cross on top of the MB Cross. Figure 4.29b shows the layout of the atmosphere side of the D-sub feedthrough.

Figure 4.28: MB Tee and PS1 Feedthrough Reducer. Left: MB Tee screwed to the MB Reducer, encasing PS1 Rear Assembly. Right: PS1 Feedthrough Reducer screwed to the MB Tee and the Bellows Ring.

Heat treatment

The MB Tee, the PS1 Front Tube, the PS1 Back Tube, the PS1 Cone, the PS2 Tube, PS2 Shell, the Front Ring, the Rear Ring, the Pushing Ring and the Bellows Ring were vacuum annealed for 48 h at 400 °C, followed by 2 h of annealing in O_2 at a pressure of 10 mbar to lower the degassing rate [11, p.13].

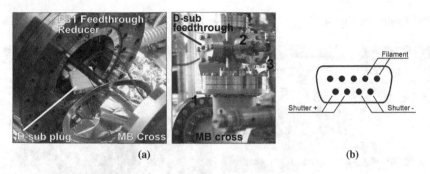

Figure 4.29: D-sub connection and D-sub feedthrough.
(a) Left: wires of the shutter and the filament are connected to a D-sub Plug. Right: CF 40 6-way cross on top of the MB Cross. 1) PS1 Feedthrough Reducer, 2) PS1 ion gauge, 3) hose to the Reservoir. The D-sub feedthrough is connected to the D-sub Plug.
(b) Layout of the atmosphere side of the D-sub feedthrough. If voltage is applied to the Shutter Motor with the shown polarity, the Shutter moves to the open position.

4.4.5 Reservoir

The inside section of the Reservoir consists of the Orifice Inset, the Reservoir Tube, the Reservoir Bellows, the Bellows Adapter and the Reservoir Feedthrough Reducer. The (non removable) Reservoir Tube Connector was inserted into the front of the Reservoir Tube, which obstructs any direct line of sight from the capillary ends to the Orifice. The Reservoir Tube Connector also holds the O-ring, which seals the Reservoir against PS1. The distance from the front of the Reservoir Tube to the Reservoir Feedthrough Reducer is approximately 80 cm.

Figure 4.30 shows the Reservoir Tube with the connected Reservoir Bellows and Bellows Adapter. The Reservoir Tube[7] is made of stainless steel (1.4404/316L), outer diameter 10 mm, inner diameter 8 mm, and is electropolished on the inside. The Reservoir Tube Connector is slid into the Orifice Inset. To monitor the temperature of the O-ring during bakeout, a K-type thermocouple is spot-welded to the front of the Reservoir Tube. The thermocouple wires are connected to a feedthrough at the CF 40 6-way cross on top of the MB Cross. The Reservoir Tube is held by the Reservoir Tube Holder, which is screwed to the inside of the

---

[7]Dockweiler Austria, Auleiten 2, A-4910 Ried im Innkreis

PS1 Cone, see Fig. 4.31.

Figure 4.30: Reservoir Tube. Top: Reservoir Tube with Reservoir Tube Holder. Left: front of Reservoir Tube with O-ring and K-type thermocouple. Right: rear end of Reservoir Tube with Reservoir Bellows and Bellows Adapter.

In the back, the Reservoir Tube is connected to the Reservoir Bellows. The Reservoir Bellows is connected to the Bellows Adapter, which has a sealing edge on its backside. Between the sealing edge and the Reservoir Feedthrough Reducer a custom-made gasket made of oxygen-free copper is placed, see Fig. 4.32. Custom-made oxygen-free copper washers are placed beneath the screw heads of the six screws pulling the Bellows Adapter against the Reservoir Feedthrough Reducer, to seal between PS1 (Bellows Adapter side) and the Reservoir (screw head side).

A standard CF 40 to CF 16 zero length reducer is connected to the backside of the Reservoir Feedthrough Reducer, which provides a standard CF 16 connection for the outside section of the Reservoir.

4.4.6 Pumping speed

Each pumping stage is pumped by a TMP with a pumping speed of $S_{N_2} = 260\,1/s$. The effective pumping speed at the volume behind Aperture 1 and Aperture 2 depends on the conductance of the tubing to the TMP. The tubing is approximated

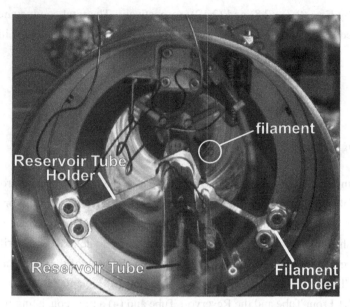

Figure 4.31: View inside the PS1 Back Tube towards the MB Core Part. The Reservoir Tube is fixed to the PS1 Cone by the Reservoir Tube Holder. At the far end of the Reservoir Tube the Orifice Inset can be seen.

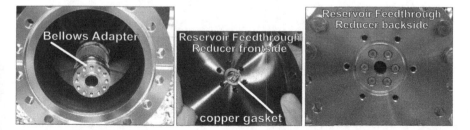

Figure 4.32: Reservoir connection at the backside of the MB Cross. Left: backside flange of MB Cross and Bellows Adapter. Middle: front side of the Reservoir Feedthrough Reducer with custom-made copper gasket (held temporarily by a piece of aluminium foil for installation). Right: backside of the mounted Reservoir Feedthrough Reducer.

by a series of orifices, tubes and annular tubes, the conductances of which are calculated by Eqs. (2.7), (2.8) and (2.12), respectively. The total conductance, C, is calculated by Eq. (2.17), for the conductance of connected components in molecular flow. The effective pumping speed, S_{eff}, is then obtained by

$$\frac{1}{S_{eff}} = \frac{1}{S_{N_2}} + \frac{1}{C}. \tag{4.51}$$

Pumping Stage 1

The tubing from the volume behind Aperture 1 to the PS1 TMP is approximated by the following components: MB Core Part (see Fig. 4.33): (1) the orifice-like area with diameter 33 mm at the back of Aperture 1 and (2) the three sections (one section shaded in Fig. 4.33) around the Orifice Inset modelled as one orifice with reduced transmission probability. The path from the back of the MB Core Part to the PS1 TMP is approximated by (see Fig. 4.14): (3) the annular tube consisting of the PS1 Front Tube and the Reservoir Tube and (4) a tube consisting of the PS1 Back Tube, the PS1 Rear Assembly and the MB Cross. The PS1 Cone between component no. 3 and no. 4 is approximated by one length of the conical transition added to the length of component no. 3. The 90° bend in the MB cross towards the TMP is already considered by one tube diameter added to the length of component no. 4. Table 4.6 summarizes the components of the approximation and their properties.

No.	Type	Properties
1	orifice	$d = 3.3\,\text{cm}$, $P = 1.0$
2	orifice	$A_{total} = 18.4\,\text{cm}^2$, $P = 0.62$
3	annular tube	$d_o = 5.0\,\text{cm}$, $d_i = 1.0\,\text{cm}$, $l = 21.7\,\text{cm}$
4	tube	$d = 9.56\,\text{cm}$, $l = 66.3\,\text{cm}$

Table 4.6: Properties of the components of the approximated tubing from the volume behind Aperture 1 to the PS1 TMP.

The conductance from the volume behind Aperture 1 to the PS1 TMP is estimated to be $C_{PS1} = 26.7\,\text{l/s}$, giving the effective pumping speed of $S_{PS1} = 24.2\,\text{l/s}$,

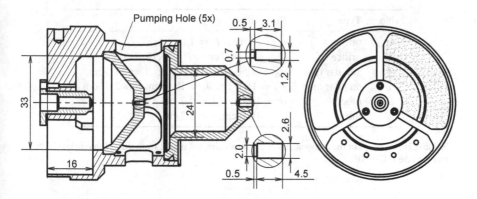

Figure 4.33: Left: cross section of the MB Core Part with mounted Orifice Inset, Aperture 1 and Aperture 2. The dimensions of Aperture 1 and Aperture 2 used in the conductance calculation are shown in detail views. Right: Rear view of the MB Core Part.

according to Eq. (4.51).

Pumping Stage 2

The tubing from the volume behind Aperture 2 to the PS2 TMP is approximated by the following components: MB Core Part (see Fig. 4.33): (1) the orifice-like area at the end of the tube-like section of Aperture 2 with diameter 24 mm (see Fig. 4.33); (2) the five pumping holes in the MB Core Part, modelled as one orifice with area A_{total} and reduced transmission probability. The path from the MB Core Part to the PS2 TMP is approximated by (see Fig. 4.14): (3) the annular tube to the MB Reducer consisting of the PS2 Shell and the PS1 Front Tube; (4) the annular tube from the MB Reducer to the height of the TMP flange of the MB Tee; (5) the tube from component no. 4 to the flange of the MB Tee and (6) the reducer from CF 150 to CF 100 modelled as a tube. The 90° bend towards the TMP is already considered by adding one outer diameter to the length of component no. 4. Table 4.7 summarizes the components of the approximation and their properties.

The conductance from the volume behind Aperture 2 to the PS2 TMP is estimated to be $C_{PS2} = 30.9\,l/s$, giving the effective pumping speed of $S_{PS2} = 27.6\,l/s$,

No.	Type	Properties
1	orifice	$d = 24\,\text{mm}$, $P = 1.0$
2	orifice	$A_{\text{total}} = 11.7\,\text{cm}^2$, $P = 0.78$
3	annular tube	$d_o = 9.56\,\text{cm}$, $d_i = 5.3\,\text{cm}$, $l = 24.3\,\text{cm}$
4	annular tube	$d_o = 15.0\,\text{cm}$, $d_i = 10.5\,\text{cm}$, $l = 25.3\,\text{cm}$
5	tube	$d = 15.0\,\text{cm}$, $l = 9.2\,\text{cm}$
6	tube	$d = 10.0\,\text{cm}$, $l = 2.2\,\text{cm}$

Table 4.7: Properties of the components of the approximated tubing from the volume behind Aperture 2 to the PS2 TMP.

according to Eq. (4.51).

4.4.7 Aperture and Shutter conductances

The conductance between PS1 and PS2 through Aperture 1, C_{a1}, and between PS2 and the chamber through Aperture 2, C_{a2}, is calculated according to Eq. (2.17), Page 11, for the conductance of connected components in molecular flow.

Aperture 1

Table 4.8 lists the components of the model for the Aperture 1 conductance: (1) the actual aperture modelled as a tube and (2) the following short tube (see Fig. 4.33). The conductance is $C_{a1} = 18.6\,\text{cm}^3/\text{s}$.

No.	Type	Dimensions
1	tube	$d = d_{a1} = 0.7\,\text{mm}$, $l = 0.5\,\text{mm}$
2	tube	$d = 1.2\,\text{mm}$, $l = 3.1\,\text{mm}$

Table 4.8: Components of the model for the Aperture 1 conductance, C_{a1}.

Aperture 2

Table 4.9 lists the components of the model for the Aperture 2 conductance: (1) the actual aperture modelled as a tube and (2) the following short tube (see Fig. 4.33). The conductance is $C_{a2} = 171.2\,\text{cm}^3/\text{s}$.

No.	Type	Dimensions
1	tube	$d = d_{a2} = 2.0\,\text{mm}, l = 0.5\,\text{mm}$
2	tube	$d = 2.6\,\text{mm}, l = 4.5\,\text{mm}$

Table 4.9: Components of the model for the Aperture 2 conductance, C_{a2}.

Shutter

When the Shutter is closed, it covers the exit of Aperture 1, which can be modelled as a circular plane with a diameter of 5 mm and hole in its center with a diameter of 1.2 mm. When the Shutter is closed, the plane and the hole are covered completely by the Shutter in an estimated maximum distance of 50 µm (see Section 4.4.2).

The conductance-defining gap between the Shutter and the plane is approximated by a narrow slit with a height of 50 µm, the conductance of which can be expressed by the molecular orifice conductance of its entrance area, $C_{0,A}$, and its transmission probability, P. The entrance area, A, is the circumference of the hole in the exit of Aperture 1 times the height of the slit, $A = 1.2\,\text{mm} \cdot \pi \cdot 50\,\mu\text{m} = 0.19\,\text{mm}^2$. The transmission probability of a slit, the height of which, b, is small compared to its width, a, with length l is given by [13, p.146]

$$P = \frac{1 + \ln(0.433 \cdot l/b + 1)}{l/b + 1}. \tag{4.52}$$

For $b = 50\,\mu\text{m}$ and a conservatively estimated "length" of the slit of $l = 2\,\text{mm}$ the transmission probability, P, is approximately 0.1. The conductance of the Shutter in closed position, C_{Shutter}, is then given by

$$C_{\text{Shutter}} = C_{O,A} \cdot P = 22.5 \, \text{cm}^3/\text{s} \cdot 0.1 = 2.3 \, \text{cm}^3/\text{s} . \qquad (4.53)$$

That suggests that the conductance between PS1 and PS2 through Aperture 1 when the Shutter is open, $C_{\text{a1}} = 18.6 \, \text{cm}^3/\text{s}$, is lowered by a factor of 0.12 when closing the Shutter; therefore, the pressure increase in PS2, $p_{2\text{i}}$ (Eq. (4.38), Page 46), should also be reduced by roughly one order of magnitude when closing the Shutter.

5 Results

5.1 Test setup

The Molecular Beam was installed on a small UHV chamber for testing. Contrary to the concept (see Section 4.2, Page 27) no capillaries were used to leak gas into the Reservoir as they were not ready at the time of the test. The theoretical MB properties in this chapter were calculated with the equivalent Orifice diameter $d_o = 38\,\mu m$ of the real Orifice in use (see Section 4.4.2, Page 54), not the conceptual Orifice diameter of 50 μm.

Molecular Beam Monitor

To record the profile of the MB the Molecular Beam Monitor (MBM) was used. The MBM is an accumulation detector consisting of an ion gauge (MBM ion gauge) connected to a small detector volume. The detector volume is accessible only through a 0.5 mm wide orifice (MBM orifice). As the MBM orifice is moved into a beam of particles the pressure in the detector volume rises until the number of particles entering and leaving the detector volume per unit time are the same. The pressure displayed by the MBM ion gauge is then a measure of the pressure of the beam. The MBM ion gauge and the detector volume sit on a 2 axis manipulator allowing to scan the plane parallel to the MBM orifice. A detailed description of the design and operation of the MBM are given in Ref. [11] and Ref. [5].

During the measurements a pumping and gas source effect of the MBM ion gauge was observed. While the detector volume is being filled, ionized gas atoms are trapped on the surfaces inside the ion gauge. While the rates of trapping and re-leasing of atoms on those surfaces approaches equilibrium, the pressure reading increases for several minutes, although the beam intensity at the MBM orifice is kept constant and the time constant of the detector volume with respect to the MBM orifice is 0.53 seconds [11]. When moving the MBM orifice out of the beam, the continuing release of previously trapped particles causes a slowly de-creasing pressure reading. A discussion of the pumping speed and the gas release in ionization gauge heads can be found in Ref. [2].

Test setup

Figure 5.1 shows a schematic of the test setup. Ar was admitted to the Reservoir by a piezoelectric leak valve (Specs Piezo Leak Valve PV16). A Pirani gauge (Leybold Pirani gauge and Leybold Combitron CM-330 controller) was used to monitor the pressure in the Reservoir. A spinning rotor gauge (MKS Spinning Rotor Vacuum Gauge 2) was also connected to the Reservoir. The pressures in Pumping Stage 1, Pumping Stage 2 and the MBM pressure were monitored by micro ion gauges (Granville-Phillips 355 Micro-Ion^{TM} Gauge), the pressure in the chamber was monitored by an ion gauge (SRS nude-UHV ion gauge). A man-ual leak valve was connected to PS2 to allow dosing of Ar to PS2. The MBM was mounted perpendicular to the MB, with the MBM orifice at the same distance from the MB as the future distance between the MB and the sample in the MRS. Pump-ing Stage 1, PS2 and the chamber were pumped by three separate TMPs (Pfeif-fer HiPace® 300), the Reservoir was pumped by a Pfeiffer HiPace® 80 TMP.

Figure 5.2a shows a photograph of the test setup. The chamber consisted of a generic CF 100 tee, two custom-made nipples (CF 100 and CF 63) and a generic CF-100 6-way cross. The chamber TMP and the chamber ion gauge were mounted on the cross, the MB and the MBM were mounted on the tee. The CF 100 (CF 63) nipple was installed between the MB (the MBM) and the tee to set the MB (the MBM) to the proper distance. Figure 5.2b shows a photograph of the equipment connected to the Reservoir.

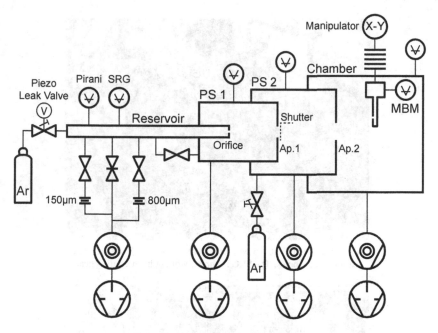

Figure 5.1: Schematic of the test setup.

The system was baked for 17 hours, after which the base pressures in PS1, PS2 and the chamber were $p_1 \approx 10^{-9}$ mbar, $p_2 < 10^{-9}$ mbar and $p_c \approx 10^{-10}$ mbar, respectively.

5.1.1 Argon correction factor

Since the Pirani gauge and the ion gauges were calibrated for N_2, the reading of the gauges had to be corrected for Ar, which was used for the testing the MB.

The Pirani gauge reading, p_{Pir}, was corrected by a factor of 1.38 for $p_{\mathrm{Pir}} \leq 6 \times 10^{-1}$ mbar, so that for the Reservoir pressure $p_{\mathrm{r}} = 1.38 \cdot p_{\mathrm{Pir}}$ was applied. The factor was obtained by comparison with the pressure reading of the SRG, see Appendix C, Page 101. For higher Pirani gauge readings the reading itself is stated.

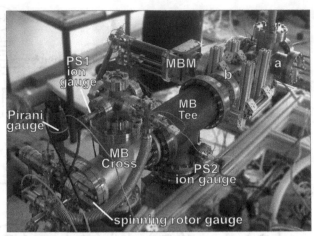

(a) a) chamber CF 100 6-way cross, b) chamber CF 100 tee and custom-made nipples. The SRG is shown without its measuring head.

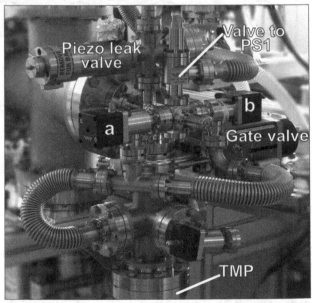

(b) Equipment connected to the Reservoir. a) 150 µm and b) 800 µm pumping orifice and valve.

Figure 5.2: Photograph of the test setup

The ion gauge readings, p_{IG}, were corrected according to the sensitivity factor of 1.29 for Ar [4, 27], so that for the stated pressures (p_1, p_2, p_c, p_{MBM}), $p = \frac{1}{1.29} \cdot p_{IG} = 0.775 \cdot p_{IG}$ was applied.

5.2 Molecular Beam profile

Ar was leaked into the Reservoir by the Piezo leak valve. The Reservoir pressure was manually stabilized while the MBM was in a retracted position, so the MBM orifice was approximately 10 cm from the MB centerline. Then the MBM was moved stepwise through the MB and the pressure reading of the MBM ion gauge, called *MBM pressure*, p_{MBM}, recorded.

The scanning direction in the line profiles was always from negative to positive horizontal positions. In the plots the horizontal position $x = 0$ was set arbitrarily for each profile.

The MBM pressure, which was read when the MBM orifice was far from the MB centerline, caused by the increased chamber pressure due to the gas introduced by the MB is called *MBM offset pressure*.

The pressure increases in PS1, PS2 and the chamber, p_{1i}, p_{2i}, p_{ci}, were obtained by subtracting the respective base pressure, p_1^b, p_2^b, p_c^b, from the current pressure reading, e.g.,

$$p_{1i} = p_1 - p_1^b . \tag{5.1}$$

5.2.1 Molecular Beam line profile

The Reservoir pressure was set to $p_r = 4.1 \times 10^{-1}$ mbar. The base pressures in PS1, PS2 and the chamber were $p_1^b = 1.4 \times 10^{-9}$ mbar, $p_2^b = 8.7 \times 10^{-9}$ mbar and $p_c^b \approx 10^{-10}$ mbar before Ar was leaked into the Reservoir.

Figure 5.3 shows p_{MBM} as a function of the horizontal position, x. The MBM offset pressure is roughly constant at $p_{MBM} = 4.0 \times 10^{-9}$ mbar. At $x = 0$ mm, the MBM orifice enters the MB core and p_{MBM} rises to 2.0×10^{-8} mbar at $x = 0.5$ mm. At $x = 3.5$ mm, the p_{MBM} reaches its maximum of 2.1×10^{-8} mbar before the MBM orifice exits the MB core and p_{MBM} decreases. The slightly increasing MBM pressure in the MB core and the "tail" after exiting the MB core is strongly believed to be caused by the pumping effect of the MBM ion gauge, see Section 5.1, Page 75.

The MB core pressure is $\tilde{p}_{MBc} = 1.7 \times 10^{-8}$ mbar (corrected for the MBM offset pressure of 0.4×10^{-8} mbar). The MB core diameter is $d_{MBc} = 3.5$ mm (from $x = 0$ to $x = 3.5$ mm) as expected. The apparent width of the MB penumbra is 0.5 mm, which is the diameter of the MBM orifice as the real "shape" of the MB is convoluted with the MBM orifice by the measurement. The PS1 penumbra and the MB Background could not be detected.

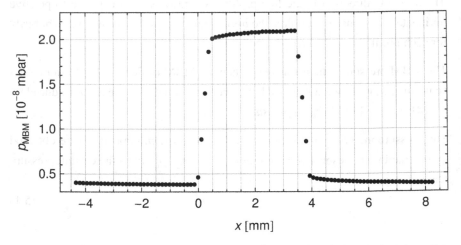

Figure 5.3: Line profile of the MB for $p_r = 4.1 \times 10^{-1}$ mbar. Shown is the MBM pressure as a function of the horizontal position, x. The horizontal position step size is 0.127 mm.

Table 5.1 shows that the MB core pressure, $\tilde{p}_{MBc}$, is in very good agreement with theory. The pressure increase in PS1, p_{1i}, is larger than the theoretical value, but still within the same order of magnitude; p_{2i} is smaller than, p_{ci} is approximately the theoretical value.

	Experiment	Theory
MB pressure [mbar]		
$\tilde{p}_{MBc}$	1.7×10^{-8}	1.9×10^{-8}
$\tilde{p}_{PS1u}$	not detectable	3.3×10^{-11}
$\tilde{p}_{MBB}$	not detectable	9.4×10^{-13}
Pressure increase [mbar]		
p_{1i}	3.1×10^{-6}	1.5×10^{-6}
p_{2i}	0.9×10^{-9}	3.0×10^{-9}
p_{ci}	$\approx 0.8 \times 10^{-10}$	0.6×10^{-10}

Table 5.1: Theoretical and experimental MB properties for a Reservoir pressure of $p_r = 4.1 \times 10^{-1}$ mbar.

Figure 5.4 shows a line profile of the rising edge (with respect to the scanning direction) of the MB, acquired in a separate measurement with decreased step size of 0.0127 mm. The line profile confirms the apparent MB penumbra width of 0.5 mm.

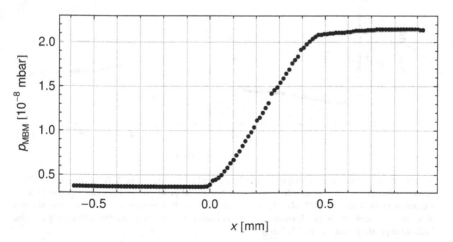

Figure 5.4: Detailed line profile of the rising edge of the MB for $p_r = 4.1 \times 10^{-1}$ mbar. Shown is the MBM pressure as a function of the horizontal position, x. The horizontal position step size is 0.0127 mm

Elevated Reservoir pressure

The Pirani gauge reading was set to $p_{Pir} = 10\,\text{mbar}$. Before Ar was leaked into the Reservoir the base pressures in PS1, PS2 and the chamber were $p_1^b = 1.6 \times 10^{-9}\,\text{mbar}$, $p_2^b = 4.7 \times 10^{-10}\,\text{mbar}$ and $p_c^b \approx 10^{-10}\,\text{mbar}$.

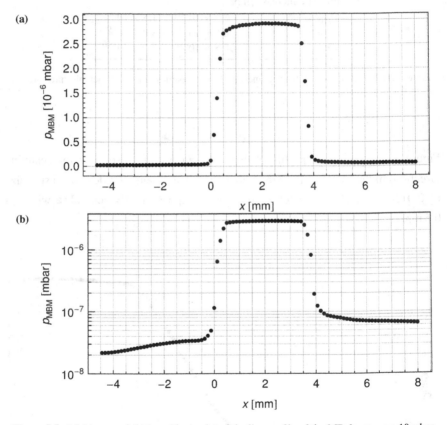

Figure 5.5: (a) Linear and (b) logarithmic plot of the line profile of the MB for $p_{Pir} = 10\,\text{mbar}$. Shown is the MBM pressure as a function of the horizontal position. The increase starting at $x \approx -0.5\,\text{mm}$ in the order of $10^{-8}\,\text{mbar}$ is associated with the PS1 penumbra. The increase starting at $x = -4.5\,\text{mm}$ and being $3.5\,\text{mm}$ in width is caused by scattering of the MB core gas. The horizontal position step size is $0.127\,\text{mm}$

Figure 5.5a shows p_{MBM} as a function of the horizontal position. The MBM offset pressure is negligibly low ($p_{MBM} = 3.0 \times 10^{-8}\,\text{mbar}$) outside of the MB core. At $x = 0$ mm, the MBM orifice enters the MB core and p_{MBM} rises to 2.7×10^{-6} mbar at $x = 0.5$ mm. At $x = 2.5$ mm, the MBM pressure reaches its maximum of 2.9×10^{-6} mbar before the MBM orifice exits the MB core again. Again, the increasing MBM pressure in the MB core is strongly believed to be caused by the pumping effect of the MBM ion gauge, see Section 5.1, Page 75.

Figure 5.5b shows the same line profile on a logarithmic scale. The MBM pressure increase starting at $x \approx -0.5$ mm in the order of 10^{-8} mbar is associated with the PS1 penumbra, which has a theoretical width of $w_{PS1p} = 0.75$ mm. To state a definite pressure increase or width by the line profile is difficult as the PS1 penumbra merges into the MB penumbra around $x = 0$. The pressure increase starting at $x = -4.5$ mm and being 3.5 mm in width is caused by scattering of the MB core gas by the tube on which the MBM orifice is mounted and which enters the MB core earlier than the MBM orifice. The scattered gas leads to an increased local chamber pressure in the chamber which is then detected by the MBM. The local chamber pressure increases as the tube is intercepting an increasing part of the MB core during the scan. As an alternative explanation, it was considered that the Orifice had an additional, unwanted opening (e.g. a tear in the stainless steel foil, which is spot-welded to the tube). This was excluded by intercepting the MB core by a rod of the MBM, approximately 10 mm away from the MBM orifice; the same effect was observed.

Table 5.2 shows a comparison between the measured MB properties and theory. First, the measured values are compared to theory for $p_r = 10$ mbar as the correction factor of the Pirani gauge is unknown for $p_{Pir} = 10$ mbar. The theoretical MB properties are too low, indicating that the reading of the Pirani gauge is smaller than the actual Reservoir pressure. Then, a Reservoir pressure was searched for, which would match the experimental MB core pressure. The MB properties for that Reservoir pressure, $p_r = 44.5$ mbar, are shown in the last column of the table; the theoretical pressure increase in PS1, p_{1i}, fits well to the measured value; p_{2i} is higher, p_{ci} is lower than the measured value. The found value for the Reservoir pressure, $p_r = 44.5$ mbar, is also in good agreement with the gas correction curve for a different Pirani gauge [26, Fig.I-2], suggesting an actual pressure of approximately 50 mbar for a 10 mbar reading.

	Experiment $p_{\text{Pir}} = 10\,\text{mbar}$	Theory $p_\text{r} = 10\,\text{mbar}$	$p_\text{r} = 44.5\,\text{mbar}$
MB pressure [mbar]			
$\tilde{p}_{\text{MBc}}$	2.9×10^{-6}	5.6×10^{-7}	2.9×10^{-6}
$\tilde{p}_{\text{PS1u}}$	$\approx 1.0 \times 10^{-8}$	9.7×10^{-10}	5.1×10^{-9}
$\tilde{p}_{\text{MBB}}$	not detectable	2.3×10^{-11}	1.5×10^{-10}
Pressure increase [mbar]			
$p_{1\text{i}}$	3.1×10^{-4}	4.5×10^{-5}	2.4×10^{-4}
$p_{2\text{i}}$	9.3×10^{-8}	4.2×10^{-8}	2.2×10^{-7}
p_{ci}	1.8×10^{-8}	1.8×10^{-9}	9.6×10^{-9}

Table 5.2: Theoretical and experimental MB properties for $p_{\text{Pir}} = 10\,\text{mbar}$.

5.2.2 Molecular Beam 2D profile

Figure 5.6 shows p_{MBM} as a function of the horizontal position, x, and the vertical position, y, for a Reservoir pressure of $p_\text{r} = 3 \times 10^{-1}\,\text{mbar}$. The MBM offset pressure is approximately $4 \times 10^{-9}\,\text{mbar}$, the MBM pressure in the MB core is approximately $2.0 \times 10^{-8}\,\text{mbar}$. The profile confirms the expected circular shape of the MB and the apparent MB penumbra width of roughly 0.5 mm on the whole circumference of the MB core. The size and shape of the MBM orifice is indicated in the figure by a dashed circle.

5.2.3 Pumping Stage 1 core and penumbra

The Reservoir and PS1 were connected, to dose Ar through the Piezo leak valve to PS1, see the schematic of the test setup (Fig. 5.1). The pressure in PS1 was set to $p_1 = 3.1 \times 10^{-4}\,\text{mbar}$.

Figure 5.7 shows p_{MBM} as a function of the horizontal position, x. The expected PS1 core and PS1 penumbra, formed by Aperture 1 and Aperture 2 can be seen.

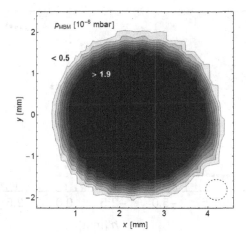

Figure 5.6: 2D profile of the MB, $p_r = 3 \times 10^{-1}$ mbar. Shown is the MBM pressure as a function of horizontal, x, and vertical position, y. The horizontal step size 0.254 mm, the vertical step size 0.2 mm, acquisition time $\approx$ 3 h. The size of the MBM orifice is indicated by the dashed circle.

The apparent width of the PS1 penumbra is approximately 1 mm, which is bigger than the theoretical value, $w_{PS1p} = 0.75$ mm, due to the convolution with the MBM orifice. The apparent overall width of approximately 5.5 mm fits well to the theoretical value of 5.5 mm, consisting of the MB core diameter, 2 times the PS1 penumbra width and the MBM orifice diameter ($d_{MBc} + 2w_{PS1p} + d_{MBM} = 5.5$ mm).

The theoretical PS1 core pressure is $\tilde{p}_{PS1u} = 6.6 \times 10^{-9}$ mbar (see Eq. (4.41), Page 47). The measured value of 3.2×10^{-9} mbar (corresponding to a MBM pressure of 8.4×10^{-9} mbar corrected for the MBM offset pressure of 5.2×10^{-9} mbar) lies within the expected order of magnitude.

5.2.4 Molecular Beam Background

Ar was leaked into PS2 by the manual leak valve connected directly to PS2, see Fig. 5.1. The pressure in PS2 was set to $p_2 = 8.5 \times 10^{-5}$ mbar. The gas effusing from PS2 into the chamber is by definition the MB Background.

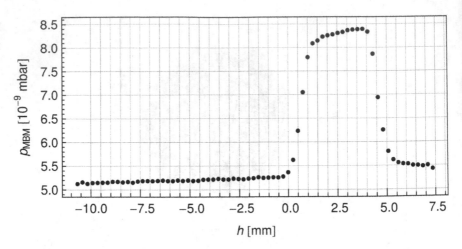

Figure 5.7: Line profile of the PS1 core and penumbra. Shown is the MBM pressure as a function of the horizontal position for $p_1 = 3.1 \times 10^{-4}$ mbar. The horizontal position step size is 0.254 mm

Figure 5.8 shows p_{MBM} as a function of the horizontal position. In the center an approximately 8 mm wide area with constant pressure is observed. Outside of this area the pressure starts to decrease due to the short tube of Aperture 2. The sudden changes of p_{MBM} at $x \approx -21$ mm and $x \approx 9$ mm were caused by unintended pressure increases in PS2. The smooth behaviour of the MB Background was confirmed in a separate measurement.

According to Eq. (4.44) the theoretical MB Background pressure in forward direction is $\tilde{p}_{MBB} = 8.0 \times 10^{-8}$ mbar. The observed $\tilde{p}_{MBB}$ of 5.8×10^{-8} mbar (MBM pressure of 2.05×10^{-7} mbar and correction for the MBM offset pressure of 1.25×10^{-7} mbar) lies within the expected order of magnitude.

5.3 Shutter

The blocking of the direct beam to the sample, the primary purpose of the shutter, was tested by operating the MB and placing the MBM orifice in the MB core center. When the shutter was closed, the MBM pressure and the chamber pressure

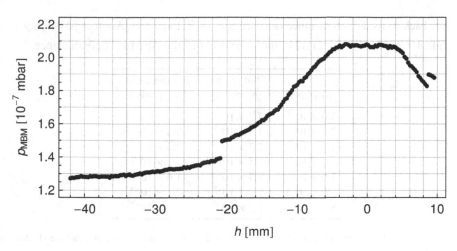

Figure 5.8: Line profile of the MB Background. Shown is the MBM pressure as a function of the horizontal position for $p_2 = 8.5 \times 10^{-5}$ mbar. The horizontal position step size is 0.254 mm. The sudden changes of p_{MBM} at $x \approx -21$ mm and $x \approx 9$ mm were caused by unintended pressure increases in PS2.

decreased quickly to their respective base pressures.

The second purpose of the shutter, decreasing the conductance of Aperture 1 between PS1 and PS2, was tested by dosing Ar to PS1 and measuring the pressure in PS2 in dependence of the shutter position. The Reservoir and PS1 were connected, to dose gas through the Piezo leak valve (located in the Reservoir) to PS1, see Fig. 5.1. First, the pressures for the open position were recorded, then the pressures for the closed position were recorded.

Table 5.3 shows the pressure set in PS1, p_1, and the measured pressure in PS2, p_2, for both shutter positions. The PS2 pressure is decreased only by 1 to 10 % by closing the shutter and not by one order of magnitude as estimated (see Section 4.4.7, Page 73). One possible explanation for the observed ineffectiveness of the shutter could be that the distance between the Shutter and the exit of Aperture 1 had increased during the assembly of the beam or the subsequent bakeout. While the shutter shaft is intended to slide freely for the compensation of possible differences in thermal expansion during bakeout, a small spring should actually move the shut-

ter back to its default distance (see Section 4.4.2, Page 59). Another possibility is
a leak between PS1 and PS2 of the same order as the Aperture 1 conductance.

p_1 [mbar]	Shutter open p_2 [mbar]	Shutter closed p_2 [mbar]
5.0×10^{-7}	7.7×10^{-10}	7.6×10^{-10}
2.0×10^{-6}	1.0×10^{-9}	9.8×10^{-10}
1.0×10^{-5}	2.3×10^{-9}	2.2×10^{-9}
5.0×10^{-5}	8.5×10^{-9}	8.3×10^{-9}
1.0×10^{-4}	1.7×10^{-8}	1.6×10^{-8}
5.0×10^{-4}	7.9×10^{-8}	7.5×10^{-8}
8.6×10^{-4}	1.3×10^{-7}	1.2×10^{-7}

Table 5.3: Pressure in PS2, p_2, in dependence of the set pressure in PS1, p_1, and the shutter position.

6 Summary and outlook

An effusive molecular beam apparatus was designed, built and tested.

The working principle of the Molecular Beam (MB) consists of a reservoir with a thin-walled orifice of 50 μm diameter as beam source. Two apertures then define the beam properties; the aperture closer to the sample defines the size of the MB core (3.5 mm) on the sample, the aperture closer to the orifice allows differential pumping of the background gas. Precise and reproducible doses to the sample are achieved by interrupting the flow of particles using a mechanical shutter. The MB allows dosing of gases, liquids and solids. Mixtures of two species can be dosed by leaking them separately to the reservoir through capillaries.

The MB was designed for a reservoir pressure of 5.4×10^{-1} mbar resulting in molecular flow conditions at the orifice and a pressure of 4.3×10^{-8} mbar in the MB core on the sample. If a higher MB core pressure is required, the reservoir pressure can be increased to 60 mbar resulting in a MB core pressure of the order 10^{-6} mbar. The MB background pressure on structures surrounding the sample is 4 orders lower than the MB core pressure, crucial for the application in temperature programmed desorption experiments.

The majority of the MB parts were produced in-house. The orifice, apertures and the MB Core Part had to be machined with very high precision and were therefore produced by an external company.

The MB was tested in a small UHV chamber. The profile of the MB was recorded with a custom-built accumulation detector [5, 11] and matches the designed profile. A MB core pressure between 1.7×10^{-8} mbar and 2.9×10^{-6} mbar was measured, that is equivalent to a dose rate between 1.3×10^{-2} L/s and 2.2 L/s. In conclusion, the apparatus works as it was designed to.

Outlook

In its current state the apparatus is ready for experiments once it is mounted on the MRS. The dosing of a mixture of two species by use of the capillaries is yet to be tested. Furthermore, the infrastructure which allows the simultaneous supply of a variety of gases, liquids and solids to the reservoir has to be developed.

A Gas properties

Gas	Standard atomic weight [8] [u]	Viscosity [17] [μPa · s]	Average velocity [Eq. (A.2)] [m/s]
H_2	2	9.0	1782
He	4	20.0	1260
H_2O	18	10.0	594
N_2	28	17.9	476
Ar	40	22.9	398
Kr	84	25.6	275
Xe	131	23.2	220

Table A.1: Standard atomic weight, viscosity and average particle velocity of gases for T = 300 K.

A.1 Maxwell-Boltzmann distribution

According to the kinetic theory of gases, particle velocities are described by the Maxwell-Boltzmann distribution [22, p.33]

$$dn\,(v) = n \cdot \left(\frac{m}{2\pi k_\mathrm{B} T} \right)^{3/2} \cdot 4\pi v^2 \cdot e^{(-mv^2/2k_\mathrm{B}T)}\, dv \,, \qquad (A.1)$$

where $dn(v)$ is the number of particles with a velocity between v and $v + dv$, m is the mass of one particle and T is the temperature. Then the most probable velocity, v_p, the average velocity, $\bar{v}$, and root mean square velocity, v_rms, are given by

$$v_{\mathrm{p}} = \sqrt{\frac{2k_{\mathrm{B}}T}{m}} \,,$$

$$\bar{v} = \sqrt{\frac{8k_{\mathrm{B}}T}{\pi m}} = \frac{2}{\sqrt{\pi}} v_{\mathrm{p}} \,, \qquad (\text{A.2})$$

$$v_{\mathrm{rms}} = \sqrt{\frac{3k_{\mathrm{B}}T}{m}} = \sqrt{\frac{3}{2}} v_{\mathrm{p}} \,.$$

B Conductance

B.1 Molecular flow in an annular tube

Several formulas for the conductance of a tube with annular cross section can be found in the literature. They will be compared to find a formula which is used in this work to calculate the conductance of an annular tube.

Knudsen's formula

Knudsen's formula gives the conductance of a tube of length L as [22, (3.92)]

$$C = \frac{8}{3\sqrt{\pi}} \sqrt{\frac{2k_B T}{m}} \frac{A^2}{BL} , \tag{B.1}$$

where A is the area and B the circumference of the cross section. For a tube with annular cross section with outer diameter D_o and inner diameter D_i, Eq. (B.1) gives [22, (3.99)]

$$C_K = \frac{1}{3} \sqrt{\frac{\pi k_B T}{2m}} \frac{(D_o^2 - D_i^2)^2}{D_o + D_i} \frac{K_0}{L} , \tag{B.2}$$

where K_0 is a an additional correction factor not included in Eq. (B.1), given in Table B.1. An adapted version of Eq. (B.2) that also includes the case of a short annular tube is given by [22, (3.114)]

$$C_{K,\text{short}} = 3.81 \sqrt{\frac{T}{M}} \frac{(D_o - D_i)^2 (D_o + D_i) K_0}{L + 1.33(D_o - D_i)} , \tag{B.3}$$

where dimensions are in cm and C in l/s, K_0 being the correction factor of Table B.1. To compute values from Eqs. (B.2) and (B.3), the values for the correction factor were interpolated by the function shown in Fig. B.1.

D_i/D_o	K_0
0	1
0.259	1.072
0.5	1.154
0.707	1.254
0.866	1.430
0.966	1.675

Table B.1: Correction factor K_0 [22, Table 3.6] for Eqs. (B.2) and (B.3).

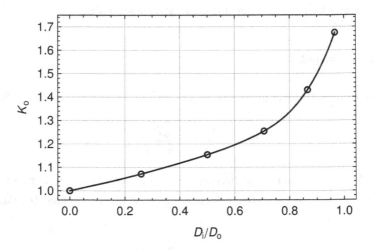

Figure B.1: Correction factor K_0 as a function of the ratio D_i/D_o. Shown are the interpolation (line) and values (circles) from Table B.1.

Measurements by Sheriff

Sherriff [25] measured conductances for tubes with $L/D_o \approx 100$ and found good agreement between conductances calculated from Eq. (B.2) without correction factor, i.e. $K_0 = 1$. Therefore Eqs. (B.2) and (B.3) without correction factor, given by

$$C_{K1} = \frac{1}{3} \sqrt{\frac{\pi k_B T}{2m}} \frac{(D_o^2 - D_i^2)^2}{D_o + D_i} \frac{1}{L}, \tag{B.4}$$

and

$$C_{K1,short} = 3.81 \sqrt{\frac{T}{M}} \frac{(D_o - D_i)^2 (D_o + D_i)}{L + 1.33(D_o - D_i)}, \tag{B.5}$$

will also be compared.

Geometric approximation

The annular tube can be approximated by similar geometries [13, p.148], depending on the ratio D_i/D_o. For $D_i/D_o \to 0$ the annular tube is approximated by a simple circular tube with $D = D_o$, and for $D_i/D_o \to 1$ by a narrow slot with height $(D_o - D_i)/2$ and width $\pi(D_i + D_o)/2$. Table B.2 shows the transmission probability for two different ratios of inner to outer diameter, read off [13, Fig. 4.39].

$2L/(D_o - D_i)$	P	$2L/(D_o - D_i)$	P
0.1	0.93	0.1	0.93
0.2	0.88	0.2	0.88
0.5	0.80	0.5	0.80
0.9	0.7	0.9	0.70
5.0	0.31	4.0	0.38
10.0	0.19	6.0	0.305
30.0	0.08	10.0	0.21
50.0	0.05	30.0	0.093
300.0	0.009	80.0	0.04
700.0	0.004	700.0	0.005
(a) Values for $D_i/D_o = 0.2$		(b) Values for $D_i/D_o = 0.8$	

Table B.2: Transmission probability values read off [13, Fig. 4.39].

Comparison and conclusion

In Figs. B.2 and B.3 the transmission probabilities for the various formulas are compared for different length to diameter ratios and inner to outer diameter ratios. C_K is only valid for long tubes, as the probability becomes larger than 1 for short tubes, which is not physical. $C_{K,short}$ shows a similar but not as bad behaviour for short tubes. $C_{K1,short}$, the formula without correction factor, behaves well for any length and inner to outer diameter ratio and is close to the geometric approximation C_{geo}. Therefore $C_{K1,short}$ is considered to be the best choice for a formula for the conductance of an annular tube:

$$C_{ann} = C_{K1,short} = 3.81 \sqrt{\frac{T}{M}} \frac{(D_o - D_i)^2 (D_o + D_i)}{L + 1.33(D_o - D_i)} , \tag{B.6}$$

where D and L are in cm and C_{ann} in l/s.

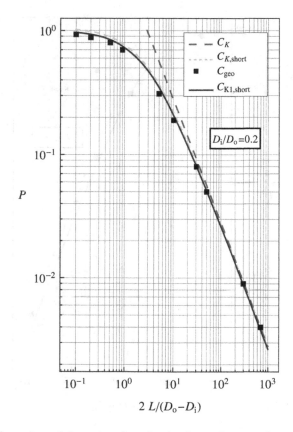

Figure B.2: Comparison of the various formulas for the conductance of annular tubes with $D_i/D_o = 0.2$. Shown is the transmission probability, P, obtained by dividing the conductance by the orifice conductance of the annular cross section, as a function of the relative length.

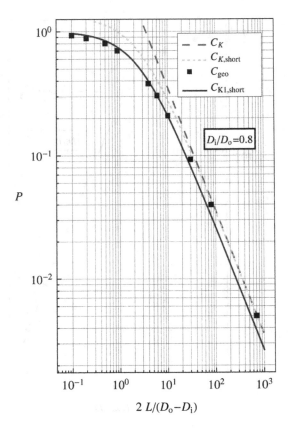

Figure B.3: Comparison of the various formulas for the conductance of annular tubes with $D_i/D_o = 0.8$. Shown is the transmission probability, P, obtained by dividing the conductance by the orifice conductance of the annular cross section, as a function of the relative length.

B.2 Orifice in transition flow

For calculation of the conductance in transition flow the measured conductance [12] for air is approximated by a linear interpolation. Table B.3 shows the experimental values read off [12, Fig.2].

Kn^{-1}	C/C_O	Kn^{-1}	C/C_O
0.1	1.0000	45.3	1.5130
0.2	1.0092	66.0	1.5221
0.3	1.0160	100.0	1.5221
0.5	1.0275	157.3	1.5313
0.7	1.0366	247.4	1.5267
1.0	1.0550	389.2	1.5221
1.6	1.0847	612.2	1.5221
2.3	1.1237	1000.0	1.5176
3.9	1.2015	1573.0	1.5130
6.6	1.2885	2474.2	1.5038
10.0	1.3618	3891.9	1.4992
14.6	1.4260	6121.9	1.4901
21.3	1.4626	10 000.0	1.4878
31.0	1.4901		

Table B.3: Experimental values for the ratio C/C_O as a function of the inverse Knudsen number, read off [12, Fig.2].

C Pirani gauge Argon correction factor

A Pirani gauge and a spinning rotor gauge (SRG) were used to monitor the reservoir pressure, p_r. Ar was dosed to the reservoir, Figure C.1 shows the SRG reading, p_{SRG}, as a function of the Pirani gauge reading, p_{Pir}. For $p_{Pir} \leq 6 \times 10^{-1}$ mbar the SRG reading is roughly linear with the Pirani gauge reading, for higher p_{Pir} the SRG reading increases more quickly. For the measured points with $p_{Pir} \leq 6 \times 10^{-1}$ mbar the average of the ratio p_{SRG}/p_{Pir} is 1.38. Therefore 1.38 is used as the Ar correction factor for the Pirani gauge, therefore $p_r = 1.38 \cdot p_{Pir}$ for a Pirani gauge reading of $p_{Pir} \leq 6 \times 10^{-1}$ mbar was applied.

p_{Pir} [mbar]	p_{SRG} [mbar]	p_{SRG}/p_{Pir}
4.0×10^{-3}	5.55×10^{-3}	1.39
6.0×10^{-3}	8.79×10^{-3}	1.47
8.0×10^{-3}	1.22×10^{-2}	1.53
1.0×10^{-2}	1.45×10^{-2}	1.45
2.0×10^{-2}	2.73×10^{-2}	1.37
4.0×10^{-2}	5.38×10^{-2}	1.35
6.0×10^{-2}	7.72×10^{-2}	1.29
8.0×10^{-2}	1.05×10^{-1}	1.31
1.0×10^{-1}	1.38×10^{-1}	1.38
2.0×10^{-1}	2.76×10^{-1}	1.38
4.0×10^{-1}	4.98×10^{-1}	1.25
6.0×10^{-1}	8.44×10^{-1}	1.41
8.0×10^{-1}	1.31	1.64
1.0	2.03	2.03

Table C.1: Measured SRG and Pirani gauge pressures for Ar.

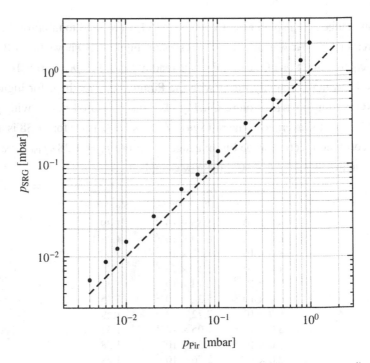

Figure C.1: Pressure reading of the SRG, p_{SRG}, as a function of the Pirani gauge reading, p_{Pir}. Dots are measured points, the dashed line was drawn to guide the eye and shows $p_{SRG} = p_{Pir}$.

References

[1] Gary Attard and Colin Barnes. *Surfaces*. Oxford University Press, 1997.

[2] A Berman. Methods of pumping speed and gas release measurement in ionization gauge heads-A review. *Vacuum*, 32(8):497–508, January 1982.

[3] J. Braun, P.K. Day, J.P. Toennies, G. Witte, and E. Neher. Micrometer-sized nozzles and skimmers for the production of supersonic He atom beams. *Rev. Sci. Instrum.*, 68:3001, 1997.

[4] Brooks Automation. *Granville-Phillips Series 358 Micro-Ion Vacuum Gauge Controller, Instruction Manual.*

[5] Florian Brunbauer. Molecular Beam Monitor Testing and Operation. *Project work*, 2014.

[6] Charles T. Campbell. Design considerations for simple gas dosers in surface science applications. *Journal of Vacuum Science & Technology A: Vacuum, Surfaces, and Films*, 3(2):408, March 1985.

[7] Peter Clausing. Über die Strahlformung bei der Molekularströmung. *Zeitschrift für Physik*, 1930.

[8] J. S. Coursey, D. J. Schwab, J. J. Tsai, and R. A. Dragoset. Atomic weights and isotopic compositions for all elements, 2013. http://physics.nist.gov/cgi-bin/Compositions/stand_alone.pl, accessed: 2013-12-08.

[9] P. Feulner. Simple ways to improve flash desorption measurements from single crystal surfaces. *Journal of Vacuum Science and Technology*, 17(2):662, March 1980.

[10] Weston M. Howard. Density Field for Rarefied Flow through an Orifice. *Physics of Fluids*, 4(4):521, 1961.

[11] Jan Hulva. Molecular Beam Monitor. *Erasmus report*, 2013.

[12] W. Jitschin, M. Ronzheimer, and S. Khodabakhshi. Gas flow measurement by means of orifices and Venturi tubes. *Vacuum*, 53(1-2):181–185, May 1999.

[13] Karl Jousten. *Wutz Handbuch Vakuumtechnik*. Springer Vieweg, 2013.

[14] D. E. Kuhl and R. G. Tobin. On the design of capillary and effusive gas dosers for surface science. *Review of Scientific Instruments*, 66(4):3016, 1995.

[15] J. Libuda and H. Freund. Molecular Beam Experiments on Model Catalysts. *Surface Science Reports*, 57:157–298, 2005.

[16] J. Libuda, I. Meusel, J. Hartmann, and H.-J. Freund. A molecular beam/surface spectroscopy apparatus for the study of reactions on complex model catalysts. *Review of Scientific Instruments*, 71(12):4395, 2000.

[17] David R. Lide. *CRC Handbook of Chemistry and Physics, Internet Version 2005*. CRC Press, Boca Raton, FL, 2005.

[18] Guangquan Lu. Elimination of serious artifacts in temperature programmed desorption spectroscopy. *Journal of Vacuum Science & Technology A: Vacuum, Surfaces, and Films*, 12(2):384, March 1994.

[19] CW Oatley. The flow of gas through composite systems at very low pressures. *British Journal of Applied Physics*, 15:15–19, 1957.

[20] John O'Hanlon. *A User's Guide to Vacuum Technology*. Wiley, 2003.

[21] H. Pauly. *Atom, Molecule, and Cluster Beams I*. Springer, 2000.

[22] Alexander Roth. *Vacuum Technology*. North-Holland, 1976.

[23] Giacinto Scoles. *Atomic and Molecular Beam Methods*. Oxford University Press, 1988.

[24] Felix Sharipov and Vladimir Seleznev. Data on Internal Rarefied Gas Flows. *Journal of Physical and Chemical Reference Data*, 27(3):657, 1998.

[25] E. Sherriff. Conductance of Bends and Concentric Tubes of a Vacuum System. *Journal of Scientific Instruments*, 43:2–5, 1949.

[26] Stanford Research Systems. *Gas Correction Curves for PG105 Gauges*.

[27] Stanford Research Systems. *IGC100 Ion Gauge Controller, Operating Manual and Programming Reference*.